Art of the Edge Tool

The Ferrous Metallurgy
of
New England Shipsmiths
and Toolmakers
1607 – 1882

Hand Tools in History Series

- Volume 6: Steel- and Toolmaking Strategies and Techniques before 1870
- Volume 7: Art of the Edge Tool: The Ferrous Metallurgy of New England Shipsmiths and Toolmakers from the Construction of Maine's First Ship, the Pinnace *Virginia* (1607), to 1882
- Volume 8: The Classic Period of American Toolmaking, 1827-1930
- Volume 9: An Archaeology of Tools: The Tool Collections of the Davistown Museum
- Volume 10: Registry of Maine Toolmakers
- Volume 11: Handbook for Ironmongers: A Glossary of Ferrous Metallurgy Terms: A Voyage through the Labyrinth of Steel- and Toolmaking Strategies and Techniques 2000 BC to 1950
- Volume 13: Tools Teach: An Iconography of American Hand Tools

Art of the Edge Tool
The Ferrous Metallurgy
of
New England
Shipsmiths and Toolmakers
From the Construction of Maine's First Ship, the Pinnace *Virginia* (1607), to 1882

H. G. Brack

Published in observation of the 400th anniversary of the reconstruction of the Pinnace *Virginia* at Fort Popham, Maine.

Davistown Museum
Museum Publication Series Volume 7

© Davistown Museum 2013
ISBN 0-9769153-5-9

Front cover illustrations:

Davistown Museum Historic Maritime IV 61204T17
Gouge signed HIGGINS & LIBBY PORTLAND

Photo titled "*Barkentine* Josephine - *598 tons - Joseph Clark & Son*" taken in 1874 at Waldoboro harbor. Used with permission from Mark W. Briscoe, 2005, *Merchant of the Medomak: Stories from Waldoboro Maine's Golden Years 1860 – 1910*, Waldoboro, ME: Waldoboro Historical Society, pg 33.

Back cover illustration:

Davistown Museum Historic Maritime IV 121906T2
Adz signed J P BILLINGS CLINTON MAINE

Cover design by Sett Balise

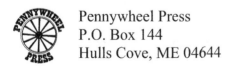

Pennywheel Press
P.O. Box 144
Hulls Cove, ME 04644

Acknowledgements

This publication was made possible by a donation from Barker Steel LLC.

Many individuals have donated to this project time, skills, information, comments, and criticisms.

I am indebted to the following: Ames Free Library-Madeline Holt, Bangor Public Library, Bridgewater Public Library-Mary S. O'Connell, Carver Public Library-Amy Sheperdson, College of William & Mary-C. R. Berquist, Jr., Concord Museum, Early American Industries Association-Elton Hall, Kingston Public Library-Carrie Elliott, Maine Historical Society, Maine Maritime Museum, Maine State Library, Maine's First Ship-Susan McChesney, Martha's Vineyard-John Flender, Martha's Vineyard Museum-Keith Gorman, Middleboro Public Library-Betty Brown, Mystic Seaport, New Bedford Whaling Museum, Norton Public Library-J. Mickelmore, Old Colony Historical Society-Jane Hennedy, Pembroke Historical Society-Debra Walt, Penobscot Maritime Museum, Richards Memorial Library-Cynthia Edson, Town of Sandwich Archives-Barbara L. Gill, Scituate Historical Society-E. Decker Adams, Scituate Town Archives-Elizabeth M. Foster, Somerset Public Library-Bonnie Mendes, Taunton Public Library-Virginia Johnson, Thomas Crane Public Library-Mary Clark, Wareham Free Library-Susan Pizzolato, Whitman Public Library-Fran Zeoli, Woods Hole Historical Collection and Museum-Susan Fletcher Witzell, Dave Brown, Rick Floyd, and Roger Smith.

Also present and endlessly helpful through all stages of the *Hand Tools in History* series and deserving of much credit for it are Judith Bradshaw Brown, Linda Dartt, Keith Goodrich, and Beth Sundberg, without whose typing, editing, filing, and patience the books would not have come to be. And thanks also to Sett Balise, whose technical skills have greatly enhanced my research and the availability of the museum's oeuvre in cyberspace.

Preface

Davistown Museum *Hand Tools in History*

One of the primary missions of the Davistown Museum is the recovery, preservation, interpretation, and display of the hand tools of the maritime culture of Maine and New England (1607-1900). The *Hand Tools in History* series, sponsored by the museum's Center for the Study of Early Tools, plays a vital role in achieving the museum mission by documenting and interpreting the history, science, and art of toolmaking. The Davistown Museum combines the *Hand Tools in History* publication series, its exhibition of hand tools, and bibliographic, library, and website resources to construct an historical overview of steel- and toolmaking strategies and techniques used by the edge toolmakers of New England's wooden age. Included in this overview are the roots of these strategies and techniques in the early Iron Age, their relationship with modern steelmaking technologies, and their culmination in the florescence of American hand tool manufacturing in the last half of the 19th century.

Background

During over 40 years of searching for New England's old woodworking tools for his Jonesport Wood Company stores, curator and series author H. G. Skip Brack collected a wide variety of different tool forms with numerous variations in metallurgical composition, many signed by their makers. The recurrent discovery of forge welded tools made in the 18th and 19th centuries provided the impetus for founding the museum and then researching and writing the *Hand Tools in History* publications. In studying the tools in the museum collection, Brack found that, in many cases, the tools seemed to contradict the popularly held belief that all shipwrights' tools and other edge tools used before the Civil War originated from Sheffield and other English tool-producing centers. In many cases, the tools that he recovered from New England tool chests and collections dating from before 1860 appeared to be American-made rather than imported from English tool-producing centers. Brack's observations and the questions that arose from them led him to research the topic and then to share his findings in the *Hand Tools in History* series.

Hand Tools in History Publications

- Volume 6: *Steel- and Toolmaking Strategies and Techniques before 1870* explores ancient and early modern steel- and toolmaking strategies and techniques, including those of early Iron Age, Roman, medieval, and Renaissance metallurgists and toolmakers. Also reviewed are the technological innovations of the Industrial Revolution, the contributions of the English industrial revolutionaries to the evolution of the factory system of mass production with interchangeable parts, and the development of bulk steelmaking processes and alloy steel technologies in the latter half of the 19th century. Many of these technologies play a role in the florescence of American ironmongers and toolmakers in the 18th and 19th centuries. Author H. G. Skip Brack cites archaeometallurgists such as Barraclough, Tylecote, Tweedle, Smith, Wertime, Wayman, and many others as useful guides for a journey through the pyrotechnics of ancient and modern metallurgy. Volume 6 includes an extensive bibliography of resources pertaining to steel- and toolmaking techniques from the early Bronze Age to the beginning of bulk-processed steel production after 1870.

- Volume 7: *Art of the Edge Tool: The Ferrous Metallurgy of New England Shipsmiths and Toolmakers* explores the evolution of tool- and steelmaking techniques by New England's shipsmiths and edge toolmakers from 1607-1882. This volume uses the construction of Maine's first ship, the pinnace *Virginia*, at Fort St. George on the Kennebec River in Maine (1607-1608), as the iconic beginning of a critically important component of colonial and early American history. While there were hundreds of small shallops and pinnaces built in North and South America by French, English, Spanish, and other explorers before 1607, the construction of the *Virginia* symbolizes the very beginning of New England's three centuries of wooden shipbuilding. This volume explores the links between the construction of the *Virginia* and the later flowering of the colonial iron industry; the relationship of 17th, 18th, and 19th century edge toolmaking techniques to the steelmaking strategies of the Renaissance; and the roots of America's indigenous iron industry in the bog iron deposits of southeastern Massachusetts and the many forges and furnaces that were built there in the early colonial period. It explores and explains this milieu, which forms the context for the productivity of New England's many shipsmiths and edge toolmakers, including the final flowering of shipbuilding in Maine in the 19th century. Also included is a bibliography of sources cited in the text.

- Volume 8: *The Classic Period of American Toolmaking 1827-1930* considers the wide variety of toolmaking industries that arose after the colonial period and its robust tradition of edge toolmaking. It discusses the origins of the florescence of American toolmaking not only in English and continental traditions, which produced gorgeous hand tools in the 18th and 19th centuries, but also in the poorly documented and often unacknowledged work of New England shipsmiths, blacksmiths, and toolmakers. This volume explicates the success of the innovative American factory system, illustrated by an ever-expanding repertoire of iron- and steelmaking strategies and the widening variety of tools produced by this factory system. It traces the vigorous growth of an American hand toolmaking industry that was based on a rapidly expanding economy, the rich natural resources of North America, and continuous westward expansion until the late 19th century. It also includes a company by company synopsis of America's most important edge toolmakers working before 1900, an extensive bibliography of sources that deal with the Industrial Revolution in America, special topic bibliographies on a variety of trades, and a timeline of the most important developments in this toolmaking florescence.

- Volume 9: *An Archaeology of Tools* contains the ever-expanding list of tools in the Davistown Museum collection, which includes important tools from many sources. The tools in the museum exhibition and school loan program that are listed in Volume 9 serve as a primary resource for information about the diversity of tool- and steelmaking strategies and techniques and the locations of manufacturers of the tools used by American artisans from the colonial period until the late 19th century.

- Volume 10: *Registry of Maine Toolmakers* fulfills an important part of the mission of the Center for the Study of Early Tools, i.e. the documentation of the Maine toolmakers and planemakers working in Maine. It includes an introductory essay on the history and social context of toolmaking in Maine; an annotated list of Maine toolmakers; a bibliography of sources of information on Maine toolmakers; and appendices on shipbuilding in Maine, the metallurgy of edge tools in the museum collection, woodworking tools of the 17th and 18th centuries, and a listing of important New England and Canadian edge toolmakers working outside of Maine. This registry is available on the Davistown Museum website and can be accessed by those wishing to research the history of Maine tools in their possession. The author greatly appreciates receiving information about as yet undocumented Maine toolmakers working before 1900.

- Volume 11: *Handbook for Ironmongers: A Glossary of Ferrous Metallurgy Terms* provides definitions pertinent to the survey of the history of ferrous metallurgy in the preceding five volumes of the *Hand Tools in History* series. The glossary defines terminology relevant to the origins and history of ferrous metallurgy, ranging from ancient metallurgical techniques to the later developments in iron and steel production in America. It also contains definitions of modern steelmaking techniques and recent research on topics such as powdered metallurgy, high resolution electron microscopy, and superplasticity. It also defines terms pertaining to the growth and uncontrolled emissions of a pyrotechnic society that manufactured the hand tools that built the machines that now produce biomass-derived consumer products and their toxic chemical byproducts. It is followed by relevant appendices, a bibliography listing sources used to compile this glossary, and a general bibliography on metallurgy. The author also acknowledges and discusses issues of language and the interpretation of terminology used by ironworkers over a period of centuries. A compilation of the many definitions related to iron and steel and their changing meanings is an important component of our survey of the history of the steel- and toolmaking strategies and techniques and the relationship of these traditions to the accomplishments of New England shipsmiths and their offspring, the edge toolmakers who made shipbuilding tools.

- Volume 13 in the *Hand Tools in History* series explores the iconography (imagery) of early American hand tools as they evolve into the Industrial Revolution's increased diversity of tool forms. The hand tools illustrated in this volume were selected from the Davistown Museum collection, most of which are cataloged in *An Archaeology of Tools* (Volume 9 in *Hand Tools in History*), and from those acquired and often sold by Liberty Tool Company and affiliated stores, collected during 40+ years of "tool picking." Also included are important tools from the private collections of Liberty Tool Company customers and Davistown Museum supporters. Beginning with tools as simple machines, reviews are provided of the metallurgy and tools used by the multitasking blacksmith, shipsmith, and other early American artisans of the Wooden Age. The development of machine-made tools and the wide variety of tool forms that characterize the American factory system of tool production are also explored. The text includes over 800 photographs and illustrations and an appendix of the tool forms depicted in Diderot's *Encyclopedia*. This survey provides a guide to the hand tools and trades that played a key role in America's industrial renaissance. The iconography of American hand tools narrates the story of a cascading series of Industrial Revolutions that culminate in the Age of Information Technology.

The *Hand Tools in History* series is an ongoing project; new information, citations, and definitions are constantly being added as they are discovered or brought to the author's attention. These updates are posted weekly on the museum website and will appear in future editions. All volumes in the *Hand Tools in History* series are available as bound soft cover editions for sale at the Davistown Museum, Liberty Tool Co., local bookstores and museums, or by order from www.davistownmuseum.org/publications.html, Amazon.com, Amazon.co.uk, CreateSpace.com, Abebooks.com, and Albris.com.

Table of Contents

Introduction

This volume of the *Hand Tools in History* series explores the stories told by the forge welded edge tools discovered in New England tool chests and workshops during the author's 40 years of searching out useful woodworking tools for the Jonesport Wood Company. *Art of the Edge Tool* examines early American toolmakers' remarkable ability to forge edge tools. It explains the milieu of these toolmakers and links it to New England's maritime trading economy and the late 19th century florescence of American hand tool manufacturing that followed.

Many of the tools recovered during the author's tool-buying expeditions were signed by their makers, who sometimes included the location of their manufacture. Often used by the shipbuilders of New England's wooden age, a majority of the larger edge tools recovered from New England workshops and collections appear to be American-made, rather than forged at Sheffield, Birmingham, or other English toolmaking centers, as many others posited. The author's observation that a robust indigenous colonial and early American edge toolmaking community clearly existed in New England, but had been poorly documented, led him to research and write this survey of the roots and evolution of New England's bloomsmiths, shipsmiths, and edge toolmakers.

A special focus of this volume is a review of the steel- and toolmaking strategies and techniques so essential to understanding the accomplishments of the forgotten shipsmiths and toolmakers who were responsible for the success of New England's maritime economy. It discusses and attempts to answer the previously unanswered questions of when, how, and where New England shipsmiths "ironed" wooden sailing ships built in New England, forged edge tools for the shipwrights who built the ships, and where they obtained their iron and steel.

Many of the small quantity of hand tools used by shipwrights and other artisans that survive from the colonial era and almost all the heavy duty shipbuilding tools used in America after the American Revolution appear to be domestically produced, not imported, edge tools. Their presence raises the following questions about the work of early shipsmiths and edge toolmakers:

- When did shipsmiths and edge toolmakers begin making edge tools instead of using those brought from England and elsewhere?
- When they began forging edge tools for colonial shipwrights, where did they obtain their steel?
- When did shipsmiths and edge toolmakers begin using steel made in the colonies or the early Republic instead of imported English and German steel to make edge tools?

The search for answers to these questions, admittedly an impossible task, forms the context for our exploration of the art of forging edge tools and the ferrous metallurgy of New England's shipsmiths and edge toolmakers.

Art of the Edge Tool: The Ferrous Metallurgy of New England Shipsmiths and Toolmakers begins with the Popham Expedition of 1607 and its attempt to build a settlement at the mouth of the Kennebec River. The construction of the pinnace *Virginia* at this location is a historic moment in New England's maritime history and an iconic event that links later colonial steel- and toolmaking strategies with earlier English and continental traditions. Colonial New England's first known shipsmith worked at this location. A description of the edge tools used by the shipwright is followed by commentary on the forgotten role of the shipsmith in colonial New England's vigorous shipbuilding industry and the historical significance of Boston as America's early colonial center of trade.

The second section of this survey reviews and summarizes steel- and toolmaking strategies and techniques from the 15th to the 19th century used by English, then American, forge masters, blacksmiths, and shipsmiths from the centuries before the settlement of North America until the era of bulk steel production (1870). These strategies and techniques were the basis for the success of New England's shipsmiths and edge toolmakers in the critical years between 1650 and 1850. The third section of the *Art of the Edge Tool* explores the growth of shipbuilding and trading in colonial New England and the roots of the American iron industry in the bog iron deposits and forges of southeastern Massachusetts. The fourth and last component comments on the great florescence of shipbuilding on Massachusetts Bay and then in Maine in the 19th century and summarizes the impact of rapidly changing technologies on Maine and New England shipbuilders and shipsmiths. An exploration of a few obscure chapters in downeast Maine history of particular familiarity and interest to the author is followed by final comments on the need to retrieve and re-narrate lost chapters in the labyrinths of our metallurgical and maritime history.

Time Line

Table 1. The following time line provides a brief summary of the chronology of important events pertaining to the steel- and toolmaking strategies of New England's shipsmiths and edge toolmakers. Descriptions with an * are excerpted from Barraclough (1984a).

Date	Event
1900 BC	First production of high quality steel edge tools by the Chalybeans from the iron sands of the south shore of the Black Sea.
1200 BC	Steel is probably being produced by the bloomery process.*
800 BC	Carburizing and quenching are being practiced in the Near East.*
800 BC	Celtic metallurgists begin making natural steel in central and eastern Europe.
650 BC	Widespread trading throughout Europe of iron currency bars, often containing a significant percentage of raw steel
400 BC	Tempered tools and evidence for the 'steeling' of iron from the Near East*
300 BC	The earliest documented use of crucibles for steel production was the smelting of Wootz steel in Muslim communities (Sherby 1995a).
200 BC	Celtic metallurgists begin supplying the Roman Republic with swords made from manganese-laced iron ores mined in Austria (Ancient Noricum).
55 BC	Julius Caesar invades Britain
50 BC	Ancient Noricum is the main center of Roman Empire ironworks. Important iron producing centers are also located in the Black Mountains of France and southern Spain.
43-410 AD	Romans control Britain.
125 AD	Steel is made in China by 'co-fusion'.*
700	High quality pattern-welded swords being produced in the upper Rhine River watershed forges by Merovingian swordsmiths from currency bars smelted in Austria and transported down the Iron Road to the Danube River.
1000	First documented forge used by the Vikings at L'Anse aux Meadows (Newfoundland)
1350	First appearance of blast furnaces in central and northern Europe
+/- 1465	First appearance of blast furnaces in the Forest of Dean (England)
1509	[Natural] steel made in the Weald [Sussex, England] by fining cast iron*
1601	First record of the cementation process, in Nuremberg*
1607	First shipsmith forge in the American colonies used at Fort St. George, Maine

Date	Event
1613/1617	Cementation process is patented in England.*
1617-1619	The great pandemic sweeps through the indigenous communities of the New England coast east of Narragansett Bay.
1625	First Maine shipsmith, James Phipps, working at Pemaquid
1629-1642	The great migration of Puritans from England brings hundreds of trained shipwrights, shipsmiths, and ironworkers to New England.
1646	First colonial blast furnaces and integrated ironworks are established at Quincy and Saugus, Massachusetts.
1652	James Leonard establishes the first of a series of southeastern Massachusetts colonial era bog iron forges on Two Mile River at Taunton, Massachusetts.
1675-1676	King Philip's War in southern Massachusetts and Rhode Island
1676	The great diaspora (scattering) of Maine residents living east of Wells follows the King Philip's War
1686	First documented use of the cementation process in England
1689-1697	The war of the League of Augsburg
1702-1714	The war of Spanish Succession
1703	Joseph Moxon ([1703] 1989) publishes *Mechanick Exercises or the Doctrine of Handy-Works*.
1709	Abraham Darby discovers how to use coke instead of coal to fuel a blast furnace.
1713	First appearance of clandestine steel cementation furnaces in the American colonies
1720	First of the Carver, Massachusetts blast furnaces established at Popes Point
+/- 1720	William Bertram invents manufacture of 'shear steel' on Tyneside.*
1722	René de Réaumur (1722) provides the first detailed European account of malleableizing cast iron.
1742	Benjamin Huntsman adapts the ancient process of crucible steel-production for his watch spring business in Sheffield, England.
1754-1763	The Seven Years War in Europe results in the last of the French and Indian wars in eastern North America
1758	John Wilkinson begins the production of engine cylinders made with the use of his recently invented boring machine.
1759	The defeat of the French at Quebec by the English signals the end of the struggle for control of eastern North America.
1763	The Treaty of Paris opens up eastern Maine for settlement by English colonists
1763-1769	James Watt designs and patents an improved version of the Newcomb atmospheric engine, i.e. the steam engine.

Date	Event
1774	John Wilkinson begins the mass production of engine cylinders used in Watt's steam engine pressure vessels.
1775	Matthew Boulton and James Watt begin mass production of steam engines.
+/-1783	The approximate date when Josiah Underhill began making edge tools in Chester, NH. The Underhill clan continued making edge tools in NH and MA until 1890
1783	James Watt improves the efficiency of the steam engine with introduction of the double-acting engine.
1784	Henry Cort introduces his redesigned reverbatory puddling furnace, allowing the decarburization of cast iron to produce wrought and malleable iron without contact with sulfur containing mineral fuels.
1784	Henry Cort invents and patents grooved rolling mills for producing bar stock and iron rod from wrought and malleable iron.
1789-1807	Era of great prosperity for New England merchants due to the neutral trade
1793	Samuel Slater begins making textiles in Pawtucket
1802-1807	Henry Maudslay invents and produces 45 different types of machines for mass production of ship's blocks for the British Navy.
1804	Samuel Lucas of Sheffield invents the process of rendering articles of cast iron malleable.
1813	Jesse Underhill is first recorded as making edge tools in Manchester, NH.
1815-1835	The factory system of using interchangeable parts for clock and gun production begins making its appearance in the United States.
1818	Thomas Blanchard designs a lathe for turning irregular gunstocks.
1820	Steam-powered saw mills come into use near Bath, Maine, shipyards.
1828	Adoption of the hot air blast improves blast furnaces
1831	Seth Boyden of Newark, NJ, first produces malleable cast iron commercially in the US.
1832	D. A. Barton begins making axes and edge tools in Rochester, NY.
1832-1853	Joseph Whitworth introduces innovations in precision measurement techniques and a standardized decimal screw thread measuring system.
1835	Malleableized cast iron is first produced in the United States.
1835	Steel is first made by the puddling process in Germany.*
1835	The first railroad is established between Boston and Worcester, Massachusetts.
1837	The Collins Axe Company in Collinsville, Connecticut, begins the production of drop-forged axes.
1837	In England, Joseph Nasmyth introduced the steam-powered rotary blowing engine.

Date	Event
1839	William Vickers of Sheffield invents the direct conversion method of making steel without using a converting furnace.
1842	Joseph Nasmyth patents his steam hammer, facilitating the industrial production of heavy equipment, such as railroad locomotives.
1849	Thomas Witherby begins the manufacture of chisels and drawknives in Millbury, MA.
1850	Joseph Dixon invents the graphite crucible used in a steel production.
1851	J. R. Brown began the manufacture of a vernier caliper
1853	John, Charles, and Richard T. Buck form the Buck Brothers Company in Rochester, NY, after emigrating from England and working for D. A. Barton. They later move to Worcester, MA in 1856 and Millbury, MA in 1864.
1856	Gasoline is first distilled at Watertown, Massachusetts.
1856	Bessemer announces his invention of a new bulk process steel-production technique at Cheltenham, England.*
1857	The panic and depression of 1857 signals the end of the great era of wooden shipbuilding in coastal New England.
1863	First successful work on the Siemens open-hearth process*
1865	Significant production of cast steel now ongoing at Pittsburg, Pennsylvania furnaces
1868	R. F. Mushet invents 'Self-hard,' the first commercial alloy steel.*
1868	Manufacture of the micrometer caliper began in America
1870-1885	Era of maximum production of Downeasters in Penobscot Bay (large four-masted bulk cargo carriers)
1874	Tilting band saw is introduced and revolutionizes shipbuilding at Essex, MA.
1879	Sidney Gilchrist Thomas invents basic steelmaking.*
1906	The first electric-arc furnace is installed in Sheffield.*
1913	Brearley invents stainless steel.*
1926	The first high-frequency induction furnace in Sheffield*

*(Barraclough 1984a, 13-4).

I. The Forgotten Shipsmith

The Building of the First Ship *Virginia*

In 1605, George Waymouth made his famous expedition to the Maine coast and explored the area around Pemaquid Point, Cushing, and the Penobscot Bay region. Two years later, Sir Ferdinando Gorges sponsored Raleigh Gilbert and George Popham's attempt to start a permanent settlement in Maine at the mouth of the Kennebec River (Ft. St. George) as part of the North Virginia Company, authorized by King James I in 1606. This settlement quickly ended as the result of a particularly brutal winter, the death of Governor George Popham, difficulties with the indigenous community of Native Americans, particularly with respect to establishing a fur trade, and Gilbert's need to return to England to settle family affairs after the death of his brother, Sir John Gilbert. Before the Popham colonists returned to England, they constructed the pinnace *Virginia*, the first vessel known to be built in continental North America and, subsequently, make multiple transatlantic voyages. The obscure but fundamental act of "ironing" this ship is the focus of our survey of the ferrous metallurgy of New England's shipsmiths and edge toolmakers. As an iconic event, it links the later steel- and toolmaking strategies and techniques used by New England ironmongers to earlier European traditions, which were the basis of the success of New England's maritime economy.

Traditionally, many of the shallops used by Basque fisherman and by Champlain and other French explorers and traders were brought to North America as breakdown kits and assembled on shore for inshore fishing, coastal exploration, or fur trading. Adverse circumstances also certainly necessitated the construction of many a shallop or pinnace in North America in the century prior to the Popham expedition. Slightly larger than the undecked single sail shallop, the pinnace *Virginia* was probably constructed entirely from wood harvested from or near the shore of the Kennebec River. The *Virginia* was built during the summer and fall of 1607 and was used to convey some of the surviving Popham colonists back to England in 1608. It returned at least once to the Maine coast after 1608 as one of several vessels in the many trading and fishing expeditions made by Sir Francis Popham, George Popham's brother, to Monhegan Island and the Maine coast in the years before John Smith's 1614 explorations. The tools used to build the *Virginia* were not produced at the site of its construction, but the existence of slag debris on private property adjoining the Fort St. George site suggests that some forging of iron bar stock used to "iron" the pinnace *Virginia* occurred at the Popham site (Bradford 2007, Brain 2007a). This iron bar stock was transported from England to America as ballast, a convenient method of transporting bar iron and other "kentledge." The fact that iron bar stock was converted to wrought iron ship fittings at Fort St. George indicates that one or more shipsmiths were already working at this location at this early date.

Among the iron artifacts recovered by Brain (2007a) at the site of the archaeological investigation of the Popham Colony were 2,431 nails or nail fragments. While these nails may have been transported from England to Fort St. George, the survival of this slag debris indicates ironworking of some type had occurred at this site. The small caulking iron that Brain found within the perimeter of the Fort St. George storehouse helps confirm the intent and ability to build a vessel such as the *Virginia*. That it might have been one of a number of irons carburized as a nest of tools by case hardening in a charcoal fire at the Popham forges is a possibility; it may also have been brought from England. The large iron washers recovered by Brain and identified as roves used to secure the spikes and planking of the *Virginia* are a more likely candidate for iron fittings produced at Fort St. George by a shipsmith during the construction of the *Virginia*.

The Popham Colony has many domains of significance for historians and history enthusiasts. Among these is Fort St. George as one of the early failed attempts to settle North America, as was the first attempt at Jamestown and the aborted settlement on the island of St. Croix (Champlain, 1604). The construction of the *Virginia* is a landmark event in the history of American shipbuilding. The obscure forge and presence of a forgotten shipsmith working at Fort St. George is an additional historical detail worthy of notice on the 400th anniversary of the failed attempt to establish the Popham Colony. The presence of a working shipsmith at the Popham settlement is a link between the Renaissance ferrous metallurgy (the smelting of iron and the manufacture of iron and steel tools) that made settlement of the New World possible and the rise of a colonial, then Federal, maritime economy in which the forgotten shipsmiths played a central role. A review of some of the basic tools manufactured by the shipsmith for the shipwright is one step in connecting the original function of the shipsmith with the rise of New England's vigorous shipbuilding industry and maritime economy.

The Edge Tools of the Shipwright

The tools used to build the pinnace *Virginia*, either as a breakdown kit from England or at the Sagadahoc settlement of Fort St. George, would have been the traditional tools of the English shipwright of the period: felling ax, broad ax, pit saw, whipsaw, frame saw, adz, slick, mast ax, mast shave, drawknife, pod auger, mortising ax, mortising chisel, and mallet. The woodworking tools illustrated in this chapter are primarily 19th century in age and are currently on exhibition at the Davistown Museum. They, nonetheless, are typical of the tools used by shipwrights in 17th and 18th century New England shipyards.

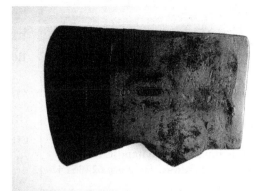

Figure 1 Yankee pattern felling ax. Forged iron and steel, 6" long with 4 ¼" blade. In the collection of the Davistown Museum TAX3500.

The first step in the construction of a wooden ship involved the felling of timber with a felling ax. Due to the large size and quantity of timber harvested in New England and elsewhere in the early colonial period, one or more unidentified toolmakers redesigned the traditional English felling ax with its light poll, creating the heavy duty welded iron and steel Yankee pattern felling ax used for two centuries of timber harvesting in New England and elsewhere.

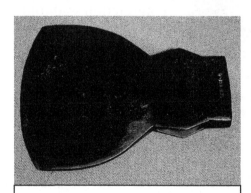

Figure 2 Forged steel New England pattern broad ax made by T. Rogers. 10 ¼" long, 7 15/16" wide. In the collection of the Davistown Museum 040904T5.

The next step in ship construction involved the use of the broad ax (Fig. 2). After felling timber for keels and futtocks, the broad ax was the principal tool used either at the felling site or in the shipyard to shape the timbers used to frame a wooden ship. Story (1995) notes that the broadax continued in use until the advent of the steam-powered band saw in 1884. "No longer need frames be beaten out with the broadaxe" (Story 1995, 113). After the timbers used for framing the ship were "beaten out" with a broad ax and transported to the shipyard, the next step was further cutting, trimming, and shaping them to construct the keel, stern, stem, futtocks, and deck frame. The frame saw (Fig. 12) was the traditional pit saw used for cutting large beams and planks in pits at least as early as the Roman Republic. In some cases, the whipsaw, a long thin flexible two man saw for

Figure 4 Yankee pattern lipped adz. Cast steel, 10 ¾" long, 2 5/8" peen, 5" wide cutting edge, 31" wood handle, signed "Collins & Co Hartford cast steel warranted legitimus" with a hammer touchmark on the handle. In the collection of the Davistown Museum 62406T4.

cutting the curved edges of a ship's keel and ribs, also was used as a pit saw. During the 18th century, the frame saw was supplanted by the "modern" pit saw, which looks like an ice saw, but is made with the lower handle attachment not used on an ice saw.

Of all the tools used by the shipwright, the most important was the adz; no wooden ship was ever constructed without the use of a stone, bronze, or steeled iron adz. The adz, a woodcutting tool with a blade at a right angle to the handle, was used to smooth the timbers cut by the broad ax and pit saw. In England, the standard English shipwrights' adz was called the peg-poll adz, a tool brought to American shipyards by immigrating English shipwrights in large quantities and copied by American shipsmiths and edge toolmakers for the next 300 years. The most common form of shipwright's adz in American shipyards after the early 19th century was the Yankee lipped adz, which, as with the American felling ax, appears

Figure 3 Peen adz, forged steel, 9" long, 3 ½" cutting edge, ¾" diameter, and signed "Thaxter Portlan_", the last letter is illegible. In the collection of the Davistown Museum 10606T1.

to be a distinctly American invention. The peg-poll adz, called a peen adz in America, continued to be used in conjunction with the lipped adz, (Figs. 3 and 4), but the lipped adz was the first preference of 19th century shipwrights, as it is still for contemporary

wooden boat builders. A fairly rare form of adz that occasionally appears in the tool chests found in maritime New England is the sharply curved gutter adz, known in England as the spoon adz. Aside from being used by house wrights to make wooden gutters, these distinctive adzes served the following special function, as noted by Horsley:

> It was designed to cut across the grain when removing large amounts of timber. Two grooves were cut with this adze, one at each end of the desired area. Then the space between could be cleared out with the normal adze. In the right hands this could be a very rapid way to remove the waste from a large joint. (Horsley 1978, 111)

Shipwrights may have used adzes for other purposes, such as hollowing out deck drains or finishing off hollowed stock used for one purpose or another. Horsley (1978) also notes another English form, the square-lipped or box adz, which appears to be a precursor to the American lipped adz. Horsley calls all of these adz forms dubbing adzes; in American terminology, "dubbing" with an adz also refers to the further shaping of the timbers used to frame a ship to eliminate the scorings left by the broad ax, such as creating a tight fit between the futtocks that composed each rib of a wooden ship. Perhaps the most well known use of the adz was for

Figure 5 Small slick, weld steel, 29 ¾" long including a 7" handle, 2 3/16" wide blade. Signed "_UGHAN & PARDO_ UNION WARRANTEED." This is Vaughan & Pardoe of Union, Maine. Working dates for this company are 1844-1868. A gift to the Davistown Museum from Rick Floyd. 21201T3.

Figure 6 Mortising Gouge, Cast steel, 13 7/8" long, 3 1/8" wide, and signed "UNDERHILL" "CAST STEEL" then upside down "BOSTON" also "TH" made with small dots. In the collection of the Davistown Museum 112303T2. It was discovered in conjunction with a Buck Brothers slick and other timber framing tools in an 18th century barn in Hanson, MA. Probably, it was once used in the heyday of shipbuilding on the North River at Scituate, Norwell, and Hanover.

smoothing out deck planking.

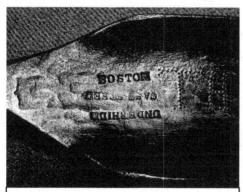

Figure 7 Detail of the mortising gouge, 112303T2, showing enlargement of the maker's mark.

The slick was another essential tool of the shipwright and served the same purpose as the adz, i.e. removing the scoring caused by the broad ax or other irregularities in the planking and frame of a wooden ship. A principal function of the slick was to smooth out wood that could not be reached with the swing of the long-handled lipped adz.

A close relative of the slick and still an occasional find in old New England tool chests is the mortising gouge (Figs. 6, 7, and on the cover). These large gouges were used for the roughing out of large mortises as noted by Horsley (1978) as the function of the spoon adz (mortising adz).

Figure 8 Cast steel, wood and brass mast shave made by Mallet, 22 ½" long, and 15 ¾" blade. In the collection of the Davistown Museum 72801T1.

A working shipwright would also use timber-framing tools and chisels of all sizes and shapes for a multiplicity of tasks. These tools ranged from small slicks and timber-framing chisels (Figs. 13 and 15), which would also be used by wharf, house, and barn builders, to slightly curved gouges, somewhat smaller than the mortising gouge, which might be used to finish off rails or spars.

The mast shave and drawknife were another important set of tools used by a shipwright. After the harvesting of timber with a felling ax and its beating out (shaping) with a broad ax, all masts and spars needed extensive additional shaping and trimming. The large drawshaves (mast shaves, Fig. 8) still occasionally found in New England tool chests attest to their use during the several centuries of wooden shipbuilding that preceded the coming of modern power tools. The mast ax is another commonplace shipyard tool with a slimmer, more streamlined design specifically used in conjunction with the mast shave to shape the mast. The ubiquitous drawshave (Fig. 9) was used on smaller wooden components of a ship for final shaping, particularly for trimming the spars that were used for booms and yards. The joiner who did interior finish work on cabins, molding, and wainscotings also used these tools, along with rabbet, fore, and joiner planes and small tools such as squares and bevels.

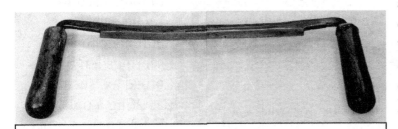

Figure 10 Drawshave, forged welded iron and steel with wooden handles, 18" long with a 11 3/8" cutting blade, signed "J.MATLACK" in the collection of the Davistown Museum 51606T6.

Figure 9 Pod auger in the collection of the Davistown Museum 070907T3.

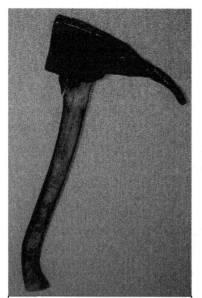

Figure 11 Mortising ax in the Collection of the Davistown Museum 102904T3. A nice example of a colonial era forge welded wrought iron and natural steel tool.

Shipwrights used mortising axes (Fig. 11) and chisels to begin the job of cutting the holes in ships' timbers, which would hold either the iron spikes and bolts or wooden treenails (trunnels) used to fasten the frame of a ship together. The pod auger (Fig. 10) was used for further shaping and enlarging these holes prior to insertion of the treenails so frequently used in the early colonial period. In the 18th and 19th century, the pod auger evolved into a variety of single and double twist, nut, and raft augers, including the useful barefoot auger, which appeared in the early 19th century.

Also of critical importance in the construction of any wooden ship, including the *Virginia*, were the caulking irons used to make the hull watertight by the insertion of oakum or tarred hemp into any spaces between the planking. This required a set of caulking irons of individual shapes for specific caulking tasks and a specially designed caulking mallet. The basic form of the caulking iron has remained unchanged since Roman times. Originally imported from Europe, examples of both English and German caulking irons were supplanted by American made irons in the late 18th or early 19th centuries. Of particular interest with respect to the construction of the pinnace *Virginia* is the recovery of a ship's caulking iron from the Popham site by Jeffrey Phipps Brain in 2004 (Fig. 16). This iron was exhibited in a show called "Popham Colony: The First English Settlement in New England 1607 – 1608" at the Maine Maritime Museum in Bath, Maine, in 2007. As noted earlier, there is no way to prove that this caulking iron was forged at the site of the slag debris found by Brain at Fort St. George; the most likely use of the forge at this location, whose existence is proven by its slag debris, was the forging of ships' fittings such as the roves collected by Brain. The repair of tools used to construct the *Virginia*, was another possible use of this forge.

Figure 12 Frame saw, forged iron and wood, 62" long, 55" long and 14 ½" wide blade, 20 ½" long handles at each end. In the collection of the Davistown Museum 1302T1.

The traditional function of the shipsmith was not only to furnish the iron fittings used to construct wooden ships but also to forge the edge tools used in ship construction. While many of the tools illustrated here are 19th century examples of the later work of New England's shipsmiths and edge toolmakers, the tradition of shipsmiths making edge tools for shipbuilding dates to the

Figure 13 Timber-framing Chisel, Cast steel, wood and iron, 2 ½" wide, 17 ¾" long. In the collection of the Davistown Museum TCC2004. Handle with forged ferrule. This tool has a distinctly forged socket. Not specifically listed in DATM but many Briggs are noted as toolmakers.

beginning of the Iron Age. Without the iron and steel tools made from the beginning of the Iron Age in Europe at Halstadt or in the Mediterranean region 400 years earlier, the evolution of long distance sea travel and trading, including the wide ranging explorations of the Phoenicians and the invasions and conquests by Roman sailors and sailing ships, would have been severely curtailed. Bronze adzes and axes built many a ship, but iron and steel edge tools were more efficient than bronze and with the availability of iron ship fittings and steeled edge tools ships grew larger and more seaworthy. The transport and establishment of early European ironworking facilities to colonial America at Saugus and elsewhere was a component of the larger phenomenon of the interrelationship between exploration and settlement of the New World and the development of sophisticated and diverse strategies for the production of iron and steel to make the edge tools used to build wooden ships. The Saugus Ironworks was the progeny of centuries of the growing finesse of European ironmongers whose skills were quickly adapted by American toolmakers and shipsmiths.

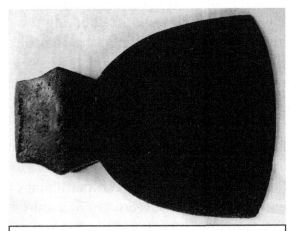

Figure 14 Broad ax, blister steel, 10" long, 6 7/8" wide blade, 3 ¼" long poll, signed "S. AVERY" and "WHORF CAST". In the collection of the Davistown Museum 41907T3.

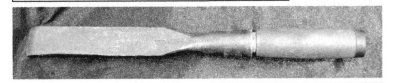

Figure 15 Socket Chisel, cast steel, 14 ¼" long including a 4 3/8" ferruled handle, signed "BILLINGS." "CAST STEEL" "CHINA" "CAST STEEL" "WARRANTED." In the collection of the Davistown Museum 81602T17. This chisel appears to be earlier than other tools made by the Billings clan except possibly John Billings of Clinton, ME (1825-1881). Did he also work in China, or is this an unrelated Billings? A previously unrecorded mark on a clearly handmade tool. Donated to the Museum by Rick Floyd.

Figure 164. Iron shipbuilding tool and hardware: a, caulking iron (P607D); b, large rove (P461D); c, small rove (P448C); d, strip of small roves (P570E). (1:1)

Figure 16 Caulking iron (a) found at the Fort St. George dig and dated to 1607-08 (Brain 2007, 140).

The Saugus Ironworks – a Link

When the Massachusetts Bay colonists came to New England during the great migration (1629 – 43), their most pressing need, after food, was ironware for shipbuilding and horticultural tools. New England's resource-based economy formed a triad: fish, timber (and furs until 1650), and shipbuilding, which necessitated steeled edge tools and ironware for ship construction. Successful survival of the communities of the early settlers was contingent upon a reliable supply of functional edge and agricultural tools. These were first imported from England, but, in 1646, one of the world's most modern integrated ironworks was constructed at Saugus, MA. A blast furnace, Walloon style finery, chafery, blacksmith's forge, and the ready availability of knowledgeable immigrant forge masters, blacksmiths, shipsmiths, and toolmakers characterized this early endeavor at ferrous metallurgy. Iron manufacturing and toolmaking on a relatively large scale had begun in North America at this facility within 26 years of the arrival of the first Plymouth colonists. The steel and toolmaking strategies and techniques available at Saugus in 1646 were unchanged from those available to the English and European blacksmiths who made the tools used to build the pinnace *Virginia*. It would be another 75 years before the first documented steel cementation furnace, which produced blister steel, would appear in the American colonies. Consideration of the strategies used to produce steel to forge the edge tools so essential for Renaissance shipwrights, Italian condottiers, or Elizabethan merchant adventurers in the years before the settlement of North America, the subject of the second section of this survey, provides the historical and technological context for the shipsmiths working in 17th century colonial New England. The evolution of blister steel production, an important technological development in steel-producing strategies, would soon impact the world of the New England shipsmiths by fostering the efficient production of high quality steel. Almost a century of vigorous shipbuilding activities in New England would pass before this technology made its appearance in the American colonies in the form of small surreptitious and often undocumented cementation steel furnaces.

Before the development of this innovation in steelmaking strategies, ancient techniques for making natural steel had been supplemented by early modern technologies associated with the use of the blast furnace and the indirect method of manufacturing wrought and malleable iron (> 1350). The Saugus Ironworks is a link between these two stages of iron and steel manufacturing and the vigorous colonial shipbuilding industry that was already well underway in New England at the time the Saugus furnace first went into blast (1646).

The role of immigrant iron smelters (bloomsmiths), blacksmiths, and shipsmiths in the colonial shipbuilding industry is poorly documented. Hartley (1957) notes the ready availability of knowledgeable ironworkers during the establishment of the first integrated ironworks at Saugus in 1646. The English Civil War (1642 – 1648) demonstrated the urgent necessity to replace British merchant ships with those owned and operated by colonial traders and merchants. The success of the "Puritan Revolution" and the beheading of James I (1649) put a temporary end to the great migration to Massachusetts Bay. The unavailability of English ships, which stopped visiting Boston because there were no more immigrants to buy their cargos, soon threatened the very existence of colonial settlements dependent on supplies and tools from England.

RICHARD E. TATREAU   *John Winthrop, Jr. Blast Furnace—August 1956*   Archaeologist ROLAND WELLS ROBBINS

Figure 17 John Winthrop's blast furnace at Furnace Brook, circa 1645, Quincy, MA (Edwards 1957).

In 1641, the Massachusetts general court initiated a program to encourage colonial construction and ownership of merchant shipping, which would assist the incipient trade with the West Indies, where lumber and fish could be sold for specie or traded for sugar. This program included the encouragement of the local iron industry and the sponsorship of John Winthrop Jr.'s successful trip to England in 1643 to obtain support and royal consent for the construction of the Saugus Ironworks (Bining 1933, Goldenberg 1976). The immediate interest of the newly arrived immigrants who settled in the Massachusetts Bay area in establishing an indigenous iron industry is made explicitly clear by the records of the Massachusetts Bay Colony, which specifically permitted and, in fact, encouraged harvesting of woodlands held in common to provide charcoal for furnaces and forges. (See Records of the Massachusetts Bay Colony, vol. 2, 1642-1649, page 81 [9/13/1644], pages 103, 104 [5/14/1645], and pages 125-128 [8/1/1645].) The intent of the colonists to mine, smelt, and forge iron and to make their own horticultural tools and shipsmiths' hardware could hardly be more clearly stated. The production of iron fittings for the booming colonial shipbuilding trade was one of the principal functions of all the first ironworks in Massachusetts (Quincy, 1645; Saugus, 1646; Taunton, 1653; Rowley, 1668) and many others during the same time period.

Unlike forges and furnaces established to the south of Boston, the Saugus Ironworks and the ironworks established at Rowley were short-lived, operating for three decades or less. Internal organizational disputes and financing may have played less of a role in the demise of the Saugus Ironworks than the rapid depletion of the limited bog iron deposits adjoining the furnace. While the Saugus Ironworks became inactive sometime in the 1670s, a network of forges and furnaces had already evolved in the bog iron district of southeastern Massachusetts, supplying the shipsmiths and blacksmiths of early colonial New England before the ironworks of Maryland, Pennsylvania, New Jersey, and elsewhere became alternative domestic sources of wrought and malleable iron bar stock after 1720.

The Colonial Legacy

Between the construction of the *Virginia* at Fort St. George in 1607-08 and the settlement of Plymouth Colony in 1620, no further documented shallop or shipbuilding occurred in New England, though construction of small vessels by shipwrecked settlers at Bermuda in 1609 and at Manhattan in 1614 is noted by Goldenberg (1976). In 1624, as narrated by William Bradford, London investors sent a ships carpenter to Plymouth who subsequently built two shallops and a cargo lighter. While constructing two ketches, "he caught a summer fever and died" (Goldenberg 1976, 7). Goldenberg notes six shipwrights arrived in Salem in 1629, and John Winthrop built his 30 ton bark *Rebecca* on the Mystic River at Medford in 1631. After this date, shipbuilding in New England, though poorly documented, became much more active due to the many shipwrights that settled in coastal communities during the great migration to Massachusetts Bay between 1629 and 1643.

In the period between 1640 and the early 20th century, large numbers of American-built ships sailed the world's oceans. Until the late 19th century, the majority of these were wooden ships built with the edge tools of the shipwright. The story of the edge tools that built the wooden ships between 1640 and the late 19th century is now nearly a forgotten chapter of our maritime history. The shipsmiths of colonial New England were artisans who used a combination of locally smelted iron and steel and imported bar iron and steel to make both the edge tools of the shipwright and the iron fittings needed for shipbuilding. They certainly did not make all the edge tools used by New England shipwrights – many continued to be imported throughout the colonial period. Enough primitive forge welded "steeled" edge tools survive from the colonial era to serve as the primary evidence of a vigorous domestic edge toolmaking community. In some cases, especially in the early colonial period, shipsmiths were both the furnace operating iron smelting forge-master who fined the bloom of iron into useful bar stock and the blacksmith who made the shipwrights' woodworking tools. They worked in the context of a diversity of strategies and techniques for making steel and steeled edge tools. Their descendents became the specialized edge toolmakers of the flowering of the New England shipbuilding industry in the middle decades of the 19th century (1840 – 1880).

New England shipsmiths had access to a unique natural resource that facilitated their tasks and helps explain the sudden flowering of New England's shipbuilding industry in the 17th century. This resource was the vast bog iron deposits of southeastern Massachusetts in Plymouth, Norfolk, and Bristol counties south of Boston and northwest of Cape Cod. These bog iron deposits were an important supplement to imported English, Swedish, Spanish, and Russian bar iron after the first blast furnaces and bloomeries were built in the Boston area (Furnace Brook, Quincy, 1645 and Hammersmith, Saugus, 1646) in the 1640s and 1650s and soon played a key role in helping local shipsmiths "iron" the many ships being built in New England.

The transition from utilizing imported, mostly English edge tools for shipbuilding to using domestically produced edge tools occurred gradually, as did the transition from relying primarily on imported English or German steel to using domestically produced steel for edge tool production. Americans did not achieve the ability to smelt, roll, and forge cast steel into high quality edge tools until the Civil War, 200 years after the development of a vigorous colonial shipbuilding industry.

During the early and mid-18th century, American (colonial) edge toolmakers introduced new variations of ancient tool forms (auger, felling ax, adz, and slick) of equal quality to English and German tools of the period. The source of steel used to produce these high quality early tools remains unknown, but domestic production of edge tools made with natural (raw) steel, blister steel (carburized wrought iron), shear steel (refined blister steel), or German steel (decarburized cast iron) facilitated the continued growth of an indigenous timber harvesting, fishing, shipbuilding, and trading economy essential for the success of the American Revolution and the post-Revolutionary War growth of the American republic.

The skills of the shipsmith as ferrous metallurgist, i.e. the maker of iron hardware and steeled edge tools, were essential to the success of the colonial and early federal economy in coasting and transoceanic trading New England. By 1776, over one third of the English merchant fleet had been constructed in the American colonies (Goldenberg 1976), and America was producing 1/7 of the world's supply of iron (Bining 1933). The individual, indigenous shipsmiths who helped construct these ships were the ironmongers living in, near, or upstream from the shipbuilding centers of New England, which eventually stretched from Passamaquoddy Bay and the Pleasant River settlements in the eastern section of the province of Maine to the southern reaches of the Connecticut coast. Beginning in the early 18th century, other shipbuilding communities emerged in the colonies to the south of New England, but they never surpassed New England in tonnage of ships built.

The bog iron deposits of southeastern Massachusetts and the forges and furnaces that smelted this iron into bar stock or cast it as hollowware and cannon balls played a critical role in the first century of New England shipbuilding. After 1720, these bog iron deposits were supplanted, then replaced by the vast rock ore deposits of western Connecticut, the Hudson River valley of New York, and especially, Maryland and Pennsylvania, and by the continued importation of foreign bar iron from Sweden and elsewhere. The milieu of

20

bloomery forges, blast furnaces, ship smithies, and rural toolmaking that this bog iron resource engendered in New England soon grew from a regional industry into a rapidly expanding colonial iron industry that provided the industrial basis for American independence.

The vigorous resource-based market economy that evolved in the colonial period laid the foundations for two revolutions: the Revolutionary War for Independence and an ongoing Industrial Revolution, which was accelerated by obscure gun-, clock-, and toolmaking innovations in the early 19th century. The tedious and now forgotten labors of two centuries of New England's shipsmiths and edge toolmakers and the bog iron deposits of southeastern Massachusetts formed the foundation of the success of the American factory system at making guns and tools with machine-made interchangeable parts and gorgeous rolled cast steel edge tools for the shipwrights of Maine and New England. Their legacy culminated with the fine tools made during the heyday of American edge tool production in the second half of the 19th century, which have never again been equaled in beauty or functional durability by domestic toolmakers.

The Role of the Shipsmith

The shipsmith is now a forgotten participant in New England's well documented maritime history. The term shipsmith seldom appears in historical surveys and local histories. It is not included in Kemp's (1976) *The Oxford Companion to Ships and the Sea* or in the voluminous index of Baker's (1973) *A Maritime History of Bath Maine and the Kennebec River Region,* where blacksmiths are noted but not toolmakers or edge tools. Complicating the issue of the role of the shipsmith, who once made the edge tools of the shipwright, is the lack of information about when and where edge toolmaking became a specialized trade. In isolated downeast Maine communities, edge toolmakers such as Benjamin Ricker (Cherryfield) were listed as shipsmiths as recently as 1856 (Parks 1857, 279).

In *Shipbuilding in Colonial America,* Goldenberg (1976, 95) notes that the expense of iron fittings used by the shipwright was approximately 16% of the cost of a typical colonial ship. The shipsmith forged these iron fittings, many of which are described below. Until sometime in the early 19th century, shipsmiths were also the most important source for the edge tools used by the shipwright. Many of the signed edge tools in the collection of The Davistown Museum dating before the era of Thomas Witherby, the Buck Brothers, and the Underhills were made by shipsmiths living near the local shipyards that they supplied with iron fittings for the shipwrights who built the ships.

The following four quotations provide the context for our search for documentation of the activities of the New England blacksmiths who were both shipsmiths and edge toolmakers. Goldenberg (1976) has compiled the most detailed record of shipbuilding in colonial America and makes the following observation about the "ironing" of a wooden ship:

> Iron was used in the superstructure of a vessel when planking was less than 1 ½ inches thick. Iron braces held the rudder to the ship, and rigging was attached to the hull with iron bolts. A 100-ton vessel would use about 1 ton of iron. This meant that a vessel such as the 300-ton ship *Mary Ann,* built at Salem in 1641, would require 3 tons of ironwork. Iron fittings became more numerous during the eighteenth century. To complicate matters, most ironwork had to be fashioned and fitted for each particular ship and therefore could not be made in England to standard sizes and then sent to the colonies. Before sailing, a vessel also had to be equipped with several anchors. (Goldenberg 1976, 16)

In his definitive chronicle of the tiny shipbuilding community of Essex in northeastern Massachusetts, Story (1995) makes this observation:

The shipsmith made all the hardware for attaching rigging to hull and spars and also made the iron fastenings used in hull construction. (Story 1995, 15)

Story also notes that the shipsmiths of Essex made the iron parts for each windlass installed on a "Chebacco" boat, the principal form of fishing schooners produced for the nearby communities of Rockport, Gloucester, and Marblehead. The shipwrights of Essex built thousands of fishing vessels for these communities and for customers in Maine and elsewhere for 250 years.

The last shipsmith to maintain a shop in Essex was Otis O. Story, who carried on into the 1930s. ...Of great significance as well was their ability to mend iron implements and tools and even, of course, to make many shipyard tools. They made many things other than tools; in 1831 Epes Story charged $1.00 for making an eel spear and 20 cents apiece for making deadeyes. (Story 1995, 122-123)

In *Material Culture of the Wooden Age,* Hindle (1981) also notes:

Almost as noisy [as the ship's caulkers] were the blacksmiths who forged the numerous iron fittings used in shipbuilding. The more careful shipwrights fastened the butts (ends) of planks with bolts instead of trunnels. Iron fittings were also used [for] deck fittings and aloft in the rigging. Another of the ironworkers was the anchorsmith, for by the eighteenth century most colonial ships carried anchors forged in America. The plentiful supply of iron in the colonies, in fact, encouraged many shipwrights to use more iron than was common in British-built ships. One builder on the North River in Massachusetts even used iron knees in place of wooden ones. Iron work totaled a little more than fifteen percent of a ship's cost in the colonial period. (Hindle 1981, 109)

In the forty years following the end of the War of 1812, wooden shipbuilding reached its peak in America. Wooden ships reached unprecedented size and swiftness, and the American merchant fleet grew far beyond its earlier numbers. Nor was the shipbuilding expansion limited to ocean-going vessels... Although this period might be called the "Golden Age" of wooden ships, those vessels contained a great deal of iron. The construction of large merchant vessels required metal bracing, tall masts supported with metal instead of hempen shrouds, and small engines on deck that hoisted sails and anchors. More metal appeared in the rigging also, and, by the Civil War, some ships carried metal masts and spars. (Hindle 1981, 122)

In his classic *The Shipwright's Trade*, Abell (1948) provides a list of the many iron fittings produced by the typical shipsmith: cables, braces, knee brackets, spikes, clinch nails, deck plate nails, penny nails, keel and futtock bolts, washers, and stern iron center lines (Abell 1948, 90). Abell also discusses the introduction of iron ships' knees, which were first produced in England and then in the United States as a result of forest depletion. The lower trunks and roots of large white oak trees growing along New England's many rivers, bays, and estuaries were the source of these ships' knees; depletion of such knees was completed in southern New England by the mid-18th century and in coastal Maine by the mid-19th century. Most knees were imported from the southern states during the 19th century, along with a wide variety of other planking and timbers, but iron knees made an occasional appearance even in New England-built ships.

Baker (1973) has an extensive discussion of the ironware needed in ship construction, noting the increasing importance of foundry-made cast iron fittings, especially after 1850, at which time the shipsmith as forger of wrought and malleable iron fittings was disappearing from many New England shipyards. The cast iron fittings of importance noted by Baker and used in Bath, Maine, area shipyards circa 1850 included: stanchion rings, side ports, side rings, hawse pipes, deck irons, chocks, quarter blocks, pump stanchions, chain sheet bushings, davit steps, and belaying pins (Baker 1973, 446). Many of these essential items had been previously made of wrought or malleable iron, as had ring bolts, chain hooks, and the ubiquitous ship spikes and bolts. Hoop iron for coopers, who played such an important role in supplying the coasting trade, was another important product of the shipsmith from the early colonial days until the era of factory-made metal hoops beginning around 1850.

By the 1830s, toolmaking techniques in America were entering a period of rapid change compared to the slow pace of change in English toolmaking centers. The American factory system of interchangeable machine-made parts was beginning to emerge; New England was its nucleus of innovation and implementation. Eli Terry had already made fundamental changes in the way clocks were made, first introducing interchangeable wooden gears, later followed by brass and iron mechanisms, which revolutionized clock making. In Portland, Maine, John Hall was beginning the design changes needed to produce guns with interchangeable parts at Harper's Ferry. Forging edge tools for the shipwright was one of the last toolmaking strategies to experience the loss of craft tradition due to mechanized production. Yet the edge toolmaking techniques of the shipsmiths and their descendents, the edge toolmaker, had been impacted by several major events in the English Industrial Revolution of the late 18th century, including the availability of expensive cast steel and cast steel tools for special purpose tasks such as carving and planing (plane irons) and the advent of rolling mills, which played such an important role in increasing the production of high quality wrought and malleable iron by squeezing out slag inclusions and standardizing bar stock production. Cort's improved rolling mills also efficiently shaped easily hot-rolled cast steel into forms compatible with the manufacture of drawshaves and timber-framing socket chisels. Yet American edge toolmakers were as conservative in their own way as the tradition-bound English toolmakers, who were so slow to adopt the revolutionary American factory system of

Edge Tool Makers.

Auster & Davis,	Addison
Avey Stephen,	Anson
Thompson M. N.,	"
Cushman Henry R.,	Andover
Finney A. R.,	Atkinson
Jackson Thomas C.,	Bath
Simpson William,	Benton
Smith Jesse & Son,	Bingham
Thomas Joseph P.,	Bluehill
Bailey Richard,	Bridgton
London Jonathan,	Bridgewater
Bussell Turner H.,	Brighton
Homer Benj. H.,	Bucksport
Homer David C.,	"
Kelly James,	Cherryfield
Lovejoy C.,	Chesterville
Breck & Weymouth,	Clinton
Woodward John,	Columbia
Morrill Joshua,	Cumberland
Dow Oliver,	Dayton
Morse J. B.,	Dixmont
Tuttle Rufus,	Durham
Seavey Charles H.,	East Machias
Seavey & Chalmer,	"
Conelon Charles R.,	"
Campbell Ambrose S.,	Ellsworth
Delanti C. L.,	"
Brown Cyrus,	Fayette
Davis Daniel,	Freeman
Durgin C. S.,	Forks
Hight Geo.,	Gorham
Morrill John J.,	Hartland
Payson John & Son,	Hope
White P.,	Jackson
Wright Joseph,	Jefferson
Bragg H. A.,	Katahdin
Cutler Levi,	Kingsbury
Doble B. W.,	Lagrange
Adams R. F.,	Lincoln
Archer J. W.,	"
Hunter David,	Lincolnville
Boothy Brice,	Limington
Davis Ezra,	Long Island
Tylor Geo. W.,	Lowell
Fairbanks Joseph,	Monmouth
White C.,	Monroe
Dennis J. M.,	New Portland
Vaughan Daniel,	New Vineyard
Mansfield E.,	Orono
Perry Henry,	Patten
Brett Rufus,	Phillips
Smart Alfred,	Pittston
Cummings Leonard F.,	Porter
Bolton E. G.,	Portland
Floyd & Stanwood,	"
Higgins & Libby,	"
Staples James,	"
Dingee & Mosher,	Presque Isle Plant.
Farnham William,	Richmond
Mallet John L.,	Rockland
Hight Amos,	Scarborough
Lyman Enoch,	Sullivan
Mowry Bradley R.,	Union
Mallet James,	Warren
White William,	Waldoboro'
Glidden Hiram,	"
Harper Samuel Jr.,	Waterboro'
Adams J. G.,	Waterford
Rounds Nathaniel,	"
North Wayne Scythe Co.,	(North) Wayne
Keyes Calvin,	Wilton
Young Joshua,	Woodstock

Figure 18 Edward C. Parks, 1857, *The Maine Register for 1857 with Business Directory for the Year 1856*. 82 Exchange Street, Portland, ME, pg. 223.

toolmaking after 1840. Maine and New England edge toolmakers, working in the tradition of the colonial shipsmith, continued to forge weld steel edge tools long after many other tools and guns were made with drop-forged interchangeable components.

The edge toolmakers of 17th and 18th century colonial America were well versed in the ancient steelmaking techniques that culminated in early modern "German" and cementation steelmaking strategies. Early colonial shipsmiths and edge toolmakers are now nearly invisible; the few edge tools made in the colonies before 1713 (the Treaty of Utrecht), when the first colonial steel furnaces appeared in New England, are impossible to assign to specific makers or forge locations. After 1720, ironworks and edge toolmaking in New England and the middle Atlantic colonies to the south grew and then flourished.

By 1750, most heavy duty edge tools used by the colonial shipwright were made in America, not in England. The tools made by shipsmiths and those made by the more specialized edge toolmakers who followed in the early 19th century are also difficult to date, their makers often remaining anonymous. However, after 1800, the vigorous market economy of the Federal Period, which arose in part due to the prosperity engendered by the neutral trade, encouraged individual toolmakers to sign their tools. Their signatures suggest that they wanted to advertise their wares for sale to other shipbuilding communities, which could easily be reached by the coasting trade. In some instances, they also noted the location of the forging of the tools they marked. The flourishing atmosphere of enterprise, exploration, and world trade resulted in craft (trade) specialization; in turn, this gave rise to the edge toolmakers of New England and elsewhere who were the descendents of the 17th and 18th century shipsmiths who first made edge tools for the colonial shipbuilding industry. The *Maine Business Directory* of 1856 (Fig. 18) expresses the growth of trade specialization that occurred in the two centuries since shipsmiths began working in New England by differentiating edge toolmaking from other trades such as blacksmithing and shipsmithing.

The many small independent edge toolmakers who worked long before the 1856 *Maine Business Directory* was published were the forebears, if not the teachers, of the later post-Civil War entrepreneurs and edge toolmakers, such as Thomas Witherby, the Underhills, and the Buck brothers, who combined advances in metallurgy, especially cast steel production, with the traditional skills of the earlier shipsmiths and edge toolmakers. The artisans who preceded these famous New England toolmakers initially made ironware and tools primarily for the shipbuilding communities in which they lived, as did the shipsmiths living in communities to the south of New England, especially in Pennsylvania, the Chesapeake Bay region, and New York. The vigorous coastwise trade soon brought New England-made tools to locations all along the Atlantic coast to supplement locally made and imported tools. In the 18th century, as New England shipsmiths perfected the art of edge toolmaking, the rapid growth of the American iron mining and smelting industry was well underway in Pennsylvania and nearby states. Important 19th century toolmaking centers soon evolved in Rochester, New York, Ohio, and elsewhere to compete with New England's many toolmakers. The flowering of America's robust 19th century gun, tool, and machinery making industries is nonetheless rooted in New England's bog iron and edge toolmaking industries. The edge tools that survive from the colonial period and the early Republic tell the story of a robust shipbuilding, iron-smelting, and toolmaking economy, without which the Declaration of Independence would have had little credibility. The milieu of the New England shipsmiths, bog iron deposits, bloomery forges, and edge toolmakers of the early Republic and their role in fostering the mid-19th century florescence of shipbuilding that followed is a poorly documented chapter in American history. A review of the beginnings and growth of shipbuilding and bloomsmithing in colonial New England helps establish the historical context for this later flowering of New England's maritime communities.

The Hidden Enablers

Following the construction of the *Virginia* at Fort St. George in 1607 – 1608, colonial shipbuilding had resumed at Plymouth as early as 1624 with the construction of two shallops for fishing and coasting and a lighter for ship to shore transport. Plymouth colonists constructed six more shallops for fishing at Salem in 1629, even before the arrival of the Winthrop fleet in 1630. Larger sailing vessels were soon constructed at Medford, Charlestown, and Boston, MA, between 1631 and 1636 (Goldenberg 1976). In 1637, William Stevens, a famed London shipwright, immigrated to Salem, helping to make that community one of New England's three early shipbuilding centers along with Boston and Charlestown. He soon moved to Marblehead and then Gloucester, and, by 1641, colonial merchant shipbuilding was well underway throughout the shipping ports north of Boston. Shallops were followed by ketches and brigantines for offshore fishing and the West Indies trade.

In 1645, another London-trained shipwright, Thomas Hawkins, built the largest ship constructed in colonial America in the 17th century, the 400 ton *Seafort*, and he followed that shortly with a 300 ton merchant vessel in 1646, both built in Boston. Nehemiah Bourne then built a 250 ton privateer (Goldenberg 1976, 12). Brought on by a shortage of specie previously supplied by immigrants in the great migration, the depression in Boston was significantly mitigated by this shipbuilding activity and the transatlantic trade in salted fish, staves, and lumber that soon resulted. Goldenberg notes that 10 large ships, smaller ketches for offshore fishing and coasting, and countless inshore fishing vessels had already been constructed in Massachusetts Bay shipyards by 1645. He makes no mention of the shipsmiths who furnished the iron fittings for these ships, but we know that the bar iron used for ships' fittings (i.e. the ironing of ships) was initially imported from England. Once the ironworks at Quincy, briefly active in 1645, and at Saugus (1646 to +/-1676), were in operation, colonial smelting of bog iron for New England's now forgotten shipsmiths was underway. After 1653, a network of bloomeries, forges, and blast furnaces were established south of Boston. Several more, such as the one at Rowley established by James and Henry Leonard, were located north of Boston. Among the most productive and long-lived forges were those established by the Leonards and Ralph Russell at Taunton in 1653. Numerous Leonards exploited the rich resources of the Taunton River watershed (Raynham, Norton, Easton, the Bridgewater, Middleboro, and Taunton) and smaller watersheds such as the Sippican River in Old Rochester (circa 1690) for the next 150 years. Bloomeries and then blast furnaces were soon established on the North River south of Braintree and later (1720) in Carver on the Weweantic, Wankinquoah (Wancinko), and Agawan rivers, the latter two of which emptied into the Wareham Narrows.

While the North River forges supplied fittings and possibly edge tools to the shipyards of Cape Cod Bay and the southern shore of the Gulf of Maine (Newburyport, Salem, and Boston) beginning in the mid-17th century, the Taunton River watershed, and to a lesser extent the Sippican and Weweantic rivers and the Wareham Narrows tidewater river watersheds, supplied ironware to the shipyards and boat shops of Buzzards and Narragansett bays from the late 17th century to the florescence of whaling ship construction in Old Rochester in the 19th century.

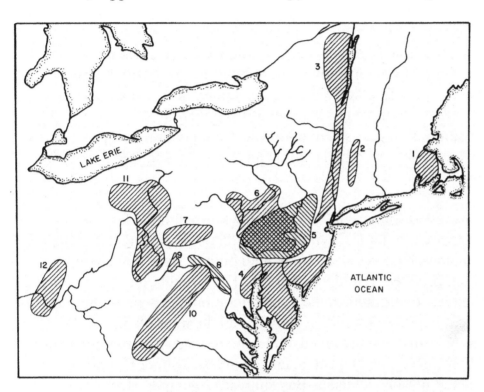

3-3. *The North American ironmaking regions in the Northeast, shown on this map, were distinguished by their natural and economic resources and by cultural preferences. Key: 1, Southeastern Massachusetts; 2, Salisbury (in Connecticut); 3, New Jersey–New York (spread along the Hudson River and Lake Champlain); 4, Coastal Plain; 5, Piedmont; 6, Anthracite (embracing the Delaware and Susquehanna Rivers); 7, Juniata; 8, Potomac; 9, Western Maryland; 10, Shenandoah (between the Potomac and James Rivers); 11, Appalachian Plateau (along the Allegheny and Monongahela Rivers); 12, Hanging Rock (in Ohio and Kentucky). Note that the piedmont and anthracite districts overlap and that the Potomac and Shenandoah districts are contiguous.*

Figure 19 Robert Gordon, 1996, *American Iron, 1607 – 1900*, Johns Hopkins University Press, Baltimore, MD, pg. 59. Region 1 on Gordon's map was the only location of significant iron-making activities in colonial America in the 17th century. Widespread mining and smelting of the rock ores in the other districts listed by Gordon occurred after 1720 and quickly superceded southeastern Massachusetts as colonial America's major region of iron production. Note the accessibility of ironworks in these regions to the west of New England to river access and therefore transport by coasting vessels.

Bining (1933) notes that many of the products of southeastern New England blast furnaces were hollowware rather than pig iron. The probability that pig iron was also cast and furnished to local blacksmiths to be fined into wrought and malleable iron for hardware and toolmaking is illustrated by the wide variety of domestically produced ironwares other than hollowware surviving from this period. As noted by

Sawyer (1988, 27, 29), in the early 19th century, blast furnaces such as that in Pocasset supplied local blacksmiths with pig iron for fining and subsequent tool production. There is no practical reason why, as at the Saugus Ironworks, blast furnaces operating after 1650 would not have supplied blacksmiths in any location with pig iron for fining into iron bar stock. The many smaller iron furnaces and forges located in southeastern New England also would have produced wrought and malleable iron directly from bog iron. Bar iron, pig iron, and steel were also imported from England, Russia, Spain, and Pennsylvania forges and furnaces. Each country or region played a role in the shipbuilding industry of New England; the strength of this industry lay in the multiple sources of steel and iron and the diversity of steel- and toolmaking strategies and techniques.

In the 18th century, the indigenous production of bog iron in southeastern Massachusetts was quickly superceded by the growing production of rock ore derived iron in the colonies to the southwest of bog iron country (Fig. 20). The primary markets for this domestically produced iron and for imported iron and steel were the growing colonial fishing, coasting, and merchant fleets, and a rapidly expanding economy that required iron tools and hardware of every description.

No edge toolmakers can be identified as making tools during the early period of colonial shipbuilding, which was centered along the New England coast from Narragansett Bay to Kittery, ME (1640 – 1720). The trading routes that evolved as a result of the Boston area shipbuilding industry were an inter-colonial network linking New England with the colonies to the south, a vigorous West Indies trade, and a transatlantic trading network that brought essential tools, bar iron, and other commodities to the colonies from Europe. In the late 18th century, the China trade expanded the scope of these trading routes after 1784, which otherwise remained unchanged despite the expanding variety of commodities being traded, until industrial innovation, westward expansion, and the spread of the railroad revolutionized the economic landscape of America.

The roots of the factory system of the 19th century lie in the triad of New England's natural resources (fish, lumber, and bog iron) and its nearly invisible enablers, the shipsmiths and shipwrights of coastal New England. Boston merchants and the London merchants, who were often their partners and the co-owners of many Boston merchant sailing vessels before and during the War of Spanish Succession, would have been unable to facilitate the expansion of colonial markets without a vigorous indigenous shipbuilding industry. This rapidly expanding market economy was dependant upon maritime trading routes that soon looped from colonial ports to the West Indies and across the Atlantic to the "wine islands" (the Canary Islands), then to Europe, Newfoundland, and back to New England.

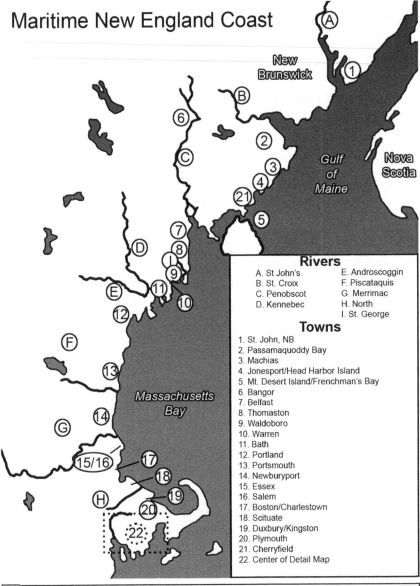

Maritime New England Coast

Rivers
A. St John's
B. St. Croix
C. Penobscot
D. Kennebec
E. Androscoggin
F. Piscataquis
G. Merrimac
H. North
I. St. George

Towns
1. St. John, NB
2. Passamaquoddy Bay
3. Machias
4. Jonesport/Head Harbor Island
5. Mt. Desert Island/Frenchman's Bay
6. Bangor
7. Belfast
8. Thomaston
9. Waldoboro
10. Warren
11. Bath
12. Portland
13. Portsmouth
14. Newburyport
15. Essex
16. Salem
17. Boston/Charlestown
18. Scituate
19. Duxbury/Kingston
20. Plymouth
21. Cherryfield
22. Center of Detail Map

Figure 20 Maritime New England coast

By 1676, Massachusetts shipwrights had constructed 430 vessels ranging in capacity from 30 to 250 tons and 300 smaller coasting and fishing vessels under 30 tons. The explosion of shipbuilding moved quickly up the New England coast to include Newbury (later Newburyport), Portsmouth, Kittery, and Wells, and south to Massachusetts at Cohasset, Scituate, Hanover, Kingston, Plymouth, and the Buzzards Bay and Narragansett Bay areas west of Cape Cod. Scituate, at the mouth of the North River, became a major shipbuilding center (1674 – 1708) and remained an important shipbuilding community for nearly two centuries despite the rapid depletion of nearby bog iron and forest resources (Goldenberg 1976, 35).

The explosion of New England shipbuilding and its expansion to areas north and south of Boston was based on the temporary availability of timber and bog iron, resources soon depleted in many locations by the rapidly expanding economic growth of the late 17th century. Major shipbuilding centers, such as Newbury, MA, and nearby towns on the Merrimac River, flourished for a century and a half primarily because thousands of square miles of timberlands lie upstream within easy reach of the Merrimac River watersheds in New Hampshire. Communities with little or no upstream resources such as Scituate or Essex, the latter of which specialized in smaller fishing vessels, imported their shipbuilding timber from Maine and the southern states until shipbuilding ended in Scituate in 1871 (Briggs 1889), or in the case of Essex, until the beginning of the 20th century.

Figure 21 Shipwright's slick, cast steel with wood handle, 14 ½" long, 3 ½" wide, 10" handle, signed "WARRANTED CAST STEEL" and "_. TINKHAM." In the Davistown Museum collection TCC2005. This slick came from a ship carpenter's tool box discovered in Foxboro, MA, several years ago and was undoubtedly used by one of the Tinkham clan, probably in the shipyards of New Bedford, Fairhaven, or Mattapoisset, MA. c. 1810 - 1850. This slick is similar to signed specimens produced by the prolific Underhill clan of Nashua, NH

The rich bog iron deposits of southeastern Massachusetts were much more extensive than those north of Boston (Saugus, Rowley) or along the North River. The vast network of forges and blast furnaces established by the Leonard family at Taunton after they left the Saugus and Braintree ironworks flourished for almost two centuries. Nearly forgotten are the shipsmiths and many later edge toolmakers, such as Levi Tinkham (Middleboro, circa 1840), who utilized bog-iron-derived bloomery iron in combination with the use of high quality imported Swedish bar iron or domestic iron from New Jersey and Pennsylvania. More well remembered are the blast furnaces and foundries of Raynham, Carver, and Bridgewater, which furnished cannon, shot, and iron sheathing for the Revolutionary War, the War of 1812, and for the U.S.S. Monitor during the Civil War. Many of the forgotten shipsmiths or their descendents of the bog iron country of southeastern Massachusetts later ironed the whaling ships built at Old Rochester (now Mattapoisett and Marion), Dartmouth, and Wareham. The identity of the whalecrafters (makers of tools for whaling) of New Bedford has been well documented (Lytle 1984; Brack 2006, pg. 282), but many other edge toolmakers and shipsmiths will forever remain unidentified.

In 1702, the port of Boston was among the most important ports in the entire British Empire. Timber for its shipwrights was brought in from all over the maritime peninsula, southern New England, and even from the central and southern coastal plain. A vigorous trade in tropical hardwoods had already developed as a component of the transatlantic trade. Rare hardwoods such as cocobolo, lignum vitae, ironwood, and more common woods, such as mahogany, were already being harvested in the Yucatan Peninsula along the Bay of Campeachy in the 17th and early 18th centuries. These tropical woods were brought to Boston and then shipped to Europe where French cabinetmakers made their contribution to an enlightenment that was always facilitated by luxury furnishings. By the mid-18th century these hardwoods were also being used by Birmingham toolmakers for the handles of chisels, drawknives, and other tools, which would often later reside in the gentlemen's toolboxes illustrated in the English pattern books that will be discussed in the next volume of the *Hand Tools in History* series. America's famous patriot-aristocrat,

Henry Knox, typifies the landed gentry who were the ultimate consumers of the finely crafted 18th century English tools, which, for so long, have helped obscure colonial America's vigorous production of more primitive tools, when beech, oak, maple, apple wood, and birch were used for the tool handles and planes of the working man.

Ships returning to Boston would import bar iron from England and Sweden to supplement the production at southeastern Massachusetts forges and furnaces, which was brought by coasting vessels to Boston, Scituate, and Newbury shipsmiths, replacing the Saugus Ironworks as a domestic source of iron after 1670. This fibrous bloomery-smelted bog iron could not supplant high quality German steel or Swedish charcoal bar iron needed for many uses including edge tool production, but it was an essential commodity for New England shipsmiths who had no need to wait for the arrival of imported Swedish iron for bolts, ship's hardware, most horticultural tools, and anchors. In addition, bloomery bog iron was higher in both phosphorus, a hardener, and silicon slag, which added corrosion resistance that was lacking in higher quality refined charcoal iron imported from integrated ironworks in Sweden, or in England from the Weald of Sussex or the Forest of Dean.

Implicit in the booming shipbuilding economy of maritime New England in the 17th century and the first decade of the 18th century were the activities of the unidentified shipsmiths who were the essential enablers of the successful settlement of colonial New England. As a matter of course, these shipsmiths would have produced iron fittings and edge tools for the shipwright; other blacksmiths produced horticultural tools and hardware of all kinds for a local market that included farmers, coopers, and other artisans. Between 1710 and 1720, radical changes occurred in the circumstances and context of the New England maritime economy. The incipient and tentative activities of the New England shipsmith, never well documented and now virtually forgotten, received a sudden stimulus for reasons having to do with international politics but not before the King Philip's War altered the southern New England landscape.

Synchronicity and Warfare

The evolution of the New England shipsmith from obscure ironmonger for the shipbuilding communities of the New England coast to sophisticated edge tool artisan of the early Republic is not a straightforward uneventful sequence. Hostilities broke out between the English and indigenous southern New England survivors of the great pandemic (1617 – 1619), the Pokanoket, who attacked Swansea, MA, in June of 1675 in response to English atrocities. This marked the beginning of 88 years of French-English conflict over the control of the North American continent. These sporadic periods of warfare also pitted indigenous communities armed with European firearms against each other in suicidal intertribal warfare and were rooted in arcane continental monarchial disputes, into which France and England were drawn as adversaries. These wars, beginning with the War of the League of Augsburg (1689 – 1697) and continuing with the War of Spanish Succession (1702 – 1714) and the last period of French and Indian warfare, the Seven Year's War (1754 – 1763), had dramatic consequences for New England shipsmiths. These wars were preceded by another war, which determined the geopolitical context for the survival and then the success of New England's maritime economy.

The short-lived King Philip's War in southern Massachusetts and Rhode Island (1675 – 1676) included the great massacre of many members of the Wampanoag community in the winter of 1676. Based on years of English – Native American resentment, violence, and cruelty, it reached its brief peak of brutality when:

> English soldiers from each of the United Colonies, aided by an Indian guide, entered the Great Swamp (near present-day South Kingston, Rhode Island), where they found a palisaded fort sheltering hundreds of wigwams and, by some estimates, as many as three thousand or four thousand Narragansetts. Most were women and children hidden in the swamp for protection during the war, along with storehouses of winter supplies. English soldiers first set fire to the wigwams and then waited as the Narragansetts began fleeing over the palisade and through its doors and windows. (Lepore 1998, 88)

The massacre that followed is now history, if only a miniscule bit of the history of European – First Nation encounters that decimated 90% of indigenous communities in all the Americas within a century of European contact.

Hostilities in coastal southeastern New England ended with the death of King Philip, son of Massassoit, in August of 1676. It directly impacted New England by opening interior and south coastal southeastern Massachusetts to English exploitation of bog iron deposits without any threat from nearby indigenous communities. When the Leonards' mills in East Taunton began producing bog iron from the lowlands of the Taunton River watershed in the 1650s, reluctant Native American communities could observe the relentless encroachment of English settlements on what had been the communal lands of First Nation peoples for millennia. The colonial forges of Taunton, the town of Middleboro, and many other communities in southeastern Massachusetts and Rhode Island were soon the target of attacks by enraged Wampanoags, Pokanokets, and Narragansetts. The great massacre at South Kingston in 1676 soon brought an end to hostilities in southern New England, opening up coastal sections of Buzzards Bay from Bourne and Wareham to Old Rochester and Old Dartmouth settlement, allowing the rapid growth of shipbuilding that began after 1680.

A curious anomaly of the King Philip's War was the time lapse between the vigorous activity of the Leonard forges in the Taunton River watershed after 1653 and the delayed appearance of significant shipbuilding activities in the Buzzards Bay shipyards of Old Rochester and those at Dartmouth and Swansea. These shipyards were established by shipbuilders from Boston, Medford, and especially, the North River, who, when they arrived between 1680 and 1720, built many of the ships used by Nantucket whalers and Martha's Vineyard merchantmen. When the King Philip's War ended, the downstream shipyards on the southeastern Massachusetts and Rhode Island coastlines began to resound with the sounds of the pit saw and broad ax, the dubbing of the adz, and the distinct ring of the caulking mallet and iron. When these sounds were heard at Fort St. George in 1607, they were the first chorus in an almost three-century-long anthem that ended only with the construction of the great downeasters in Penobscot Bay and the four-, five-, and six-masted schooners built in Waldoboro and Bath in the late 19th century. In the case of the Leonards' mills, over a quarter of a century passed before one of their ironwares market emerged in the form of the unidentified shipsmiths working at the easily accessible, if isolated, boatyards along the Buzzards Bay shoreline and in the Rhode Island archipelago of Narragansett Bay. During this interim, what were the other markets for the iron bar stock smelted at the Leonard forges?

There was no easy way to ship bog iron bar stock from the Taunton River to a shipsmith in Boston without coasting around the Nantucket Shoals. Leonards' mills and other bog iron forges in the Taunton River watershed were operating well before the establishment and operation of Thomas Coram's famous shipyard at Dighton (1697 – 1720) or any known shipbuilding activities at Taunton, Freetown, Swansea, or Rehoboth. This raises questions about the location of the markets for the bog iron produced in the Taunton River watershed, which include Middleboro, Raynham, and many other communities where bog iron was smelted at this time or shortly thereafter. Because there would have been a significant market for bog iron to supply anchor forges and shipsmiths in the many shipbuilding communities located to the north of Cape Cod, the early date of iron production in the Taunton River watershed causes one to wonder if this iron could also have been shipped overland north from Taunton to Quincy and then to Boston and whether there was a North River link requiring only a short voyage through the dense forest of Plymouth County to Pembroke, where the headwaters of the North River could be accessed? Briggs (1889) provides insight into why Scituate was such an important shipbuilding center by listing the many bog iron forges located upstream on the North River (see Appendix B.) No mention is made in Briggs or other local histories of the source of the iron used in Cape Cod shipbuilding communities.

The quick end to King Philip's War encouraged more timber and bog iron harvesting in southeastern Massachusetts, the latter of which was often done by boats with scoops in ponds now filled with cranberries. The harvesting of bog iron by families throughout the region and its sale to local forges and furnaces was a lucrative alternative source of money, or more often iron blooms for the farmyard forge, in a region where timber resources were soon depleted. In fact, the ready availability of bog iron muck bars produced by the bloomsmiths and small forges of southeastern New England provided the raw materials, wrought and malleable iron, which were then used by individual farm families who made their own tools. These hand-forged, wrought and malleable iron tools still can be found in New England in workshops and flea markets and have been well documented in Sloane's (1964) classic *A Museum of Early American Tools*. At the same time that isolated farmers were forging their own hand tools throughout colonial New England, the iron industries of southeastern Massachusetts made communities such as Taunton, Middleboro, Norton, and Bridgewater much more important colonial commercial and industrial centers, especially for ordnance production during the Revolutionary War and later during the War of 1812 than we now remember.

Important shipbuilding communities located to the north of Cape Cod, in addition to Scituate, include Hingham, Cohasset, Kingston, Duxbury, and Plymouth; all used local bog iron for the needs of the shipsmith, as did Boston, always accessible to the then vigorous coasting trade. The most productive of these Cape Cod Bay shipbuilding communities was Scituate, located at the mouth of the North River, which constructed 131 ships between 1674 and 1708, while Boston, during this same period, produced 241. This tonnage of shipbuilding would make Scituate the third largest shipbuilding community in North America, after Boston and Salem, during this time. Brigg's (1889) listing of North River bog iron forges operating in the colonial period illustrates the early reliance of New England's shipwrights on indigenous bog iron deposits. The local shipsmith would have been the best customer of these small furnaces. There would be no reason why the shipyards of Boston, Charlestown, Medford, and Salem would not also have obtained wrought iron for ship fittings from North River forges. The coastwise trade in sheltered Cape Cod Bay provided convenient inexpensive transportation for locally smelted iron bar stock, which was then forged into fittings, at or near shipyards, in the shapes, sizes, and quantities required for a particular ship design. Documentation is lacking for comparison of colonial use of imported versus locally made standard size hardware such as spikes and bolts, but it was unlikely that English ironware was cheaper or higher in quality for these easily produced items.

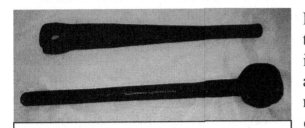

Figure 22 These shipsmith's bolt headers were made and used at a South Duxbury forge, which may have been operating as early as the late 17th century. Top: 13" long, 3/16" and 9/16" diameter round head holes, Davistown Museum 121805T3. Bottom: 12 ½" long, 11/16" wide square head, and 3/8" diameter round head hole, Davistown Museum 121805T2.

During and after the period from 1645 to 1700, the incredible productivity of the shipbuilding industry in Massachusetts mandated the availability of ironware and edge tools. There remains a question as to what extent domestically smelted bog iron supplemented imported English, and then Swedish, bar iron. Swedish iron was imported into shipbuilding communities such as Salem and Boston in large quantities throughout the late 18th century. The impost records at the New Bedford Whaling Museum detail the later importation of large quantities of Swedish iron to New Bedford from Gottenberg between 1816 and 1831; this high quality charcoal fired wrought iron was almost certainly used in cementation (steel) furnaces to produce the steel needed by the large community of New Bedford / Rochester area shipsmiths, edge toolmakers, and whalecrafters who supplied the shipwrights of New Bedford and Old Rochester during the peak years of the whaling industry. The further back in time we go, the less we know about the identity of the toolmakers and shipsmiths who helped build the fleets of the colonial era. Nonetheless, the search for the roots of New England's vigorous 19th century shipbuilding industry must begin with a consideration of the role of Boston as America's most important late 17th and early 18th

Figure 23 Slick marked B D HATHAWAY NEW BEDFORD and WARRANTED CAST STEEL. Photos taken by the Davistown Museum and used with permission of the owner Dave Brown

century port, mercantile trading center, and shipbuilding community. One hundred and fifty years before the great florescence of shipbuilding in New England in the early- and mid-19th century, in Boston and nearby coastal communities, the New England shipsmith and his edge tools were already at work building a new nation, despite the conflict between indigenous First Nation communities and European settlers for control of inland New

England that would continue until the fall of Quebec in 1759 and the Treaty of Paris in 1763.

Boston: Early Colonial Center of Trade

Goldenberg (1976) has this telling comment on Boston as an economic hub and trading center.

> In 1702 Boston was one of ten leading ports in the empire. The tonnage of Boston may be figured as second, third, or eighth to that of London, depending upon the adjustments different observers make. Even the position of eighth behind London (the calculation made by Ralph Davis) is remarkable considering Boston was the only port outside England to rate among the top ten—no Scottish, Irish, or other colonial port qualified. (Goldenberg 1976, 42)

But what lead to this hegemony of Boston as a center of trade? The answer lies in those obscure French-English conflicts, especially the War of the League of Augsburg (1689 – 1697) known as King William's War in New England and the War of the Spanish Succession (1702 – 1714), also called the Queen Anne's War. These were components of the larger French and English struggle for control of the North American continent. What possible link could there be between New England shipsmiths and continental as well as colonial warfare?

The rise of Boston as an important English port was the result of the unique circumstance of co-ownership of large numbers of transoceanic ships, about which Goldenberg (1976) and Bailyn (1959) note the following.

> Out of a total of 181 ships 100 tons or larger built between 1674 and 1714, 52 were owned entirely by Massachusetts merchants, most of whom lived in Boston and the others in Charlestown and Salem. British merchants owned 40 ships; some owners came from the West Indies, Scotland, and Bristol, but the majority were London merchants. Syndicates of Massachusetts and English merchants owned 89 ships. Altogether, British merchants had investments in 129 of the 181 large Massachusetts ships. The ratio might even be higher, for the register fails to include the licenses of vessels later registered outside Massachusetts. (Goldenberg 1976, 41)

But what led to this colonial alliance of Boston and London merchant ship oligarchs? Or, as Goldenberg asks, then explains:

> What caused this sudden growth in Massachusetts shipping? During the three brief Dutch wars of the third quarter of the seventeenth century, England had captured between 2,000 and 3,000 prizes and had lost less than 500 of its own merchant ships, but this convenient method of securing tonnage was reversed in the War of the League of Augsburg. Perhaps as many as 4,000 English vessels were captured; and all but 500 were taken in the last years of the struggle, 1694-97. These concentrated

losses resulted from the French policy of avoiding naval battles in order to pursue merchant ships. The destruction of British commerce was further hastened by increasing numbers of French privateers. English convoys tried vainly to give merchant ships better protection in the War of the Spanish Succession, as the French proceeded to capture at least 2,000 vessels. (Goldenberg 1976, 31-33)

Herein lie both the secret of the success of the New England shipsmith and the context of an ongoing fight for survival by the newly established colonial maritime communities who laid the foundations for the birth of a nation only 75 years in the future. Boston, Charlestown, Salem, and Scituate were thriving; their prosperity was intimately connected to the urgent need to replace British ships lost during the War of the League of Augsburg. While these coastal ports flourished, due in part to events in continental Europe, only a few miles to the north and west intermittent Indian attacks, encouraged and assisted by the French, were occurring at Haverhill, MA, Dover, NH, Saco, ME, and at inland Massachusetts locations, such as Lancaster, Brookfield, and many other towns. After 52 years of English occupation by nearly 10,000 settlers, King Philip's War spread to the coast of Maine, despite peace in southeastern Massachusetts, and it resulted in a second great diaspora of Maine's English coastal residents, the first having been the great pandemic of 1617-1619 and the demise of the Abenaki communities of coastal Maine. By 1676, the entire coast of Maine was cleared of all English settlements east of Wells. Hundreds of English settlers were killed or taken to the French missionary villages of the St. Lawrence River basin as captives. The existence of vigorous colonial shipbuilding activity along the shores of Massachusetts Bay, Cape Cod Bay, and Buzzards Bay is ironic, even anomalous. While Boston, Salem, and Scituate were flourishing and the colonial fishing and shipbuilding communities from Kittery to Narragansett Bay were relatively secure, the question of the successful settlement of interior New England and distant wilderness areas to the west was unresolved. Two hundred Native Americans and 45 French militia attacked Deerfield in the middle of the night on February 29th 1704; that Deerfield massacre made any idea of westward expansion or even permanent settlement of the Connecticut River Valley a remote dream. It would have been inconceivable to the fleeing residents of this area that, in just a little over a century, the Connecticut River Valley would be an important center of the growth of a domestic toolmaking industry in a new republic whose prosperity would endure until the end of the 20th century.

Cargoes and Trading Routes

> By 1665 there were three hundred New England vessels trading with Barbadoes, Virginia, Madeira, Acadia, etc., and 1,300 smaller craft were fishing at Cape Sable. Cod and mackerel were caught and salted. The best fish were sent to Malaga and the Canaries, the second sort to the Portugal Islands, and the worst to the Barbadoes there to be used in the diet of the negro slaves. At that time, the principal commodities produced in the Massachusetts Bay were fish and pipe-staves, masts, fir-boards, pitch, tar, pork, beef, and horses and corn which were sent to Virginia, Barbadoes, &c. Tobacco and sugar were taken in payment and shipped to England. (Dow 1935, 145)

Almost all of the vessels noted by Dow (1935) were built along the coast of New England. Numerous shipwrights would have been among the thousands of residents who had moved to coastal Maine prior to the Indian wars, which began in 1676. These Maine shipwrights also built some of the smaller craft and the broad beamed cordwood boats used for Maine's inshore fisheries, timber harvesting, and its incipient coasting trades. Many of the vessels built in Maine would have been too small to be included in the colonial register of shipping that Goldenberg (1976) cites. The identity of most Maine shipwrights and the location of their activities have now been lost to us. The cord-wood boats brought essential firewood to the already thriving but firewood- starved port of Boston and to neighboring Salem and other coastal communities. They would have also brought many of the products shipped out in the coasting trade, including the staves and fir boards noted by Dow, and also many other wooden products such as house frames, clapboards, trawl-line tubs, salt boxes, firkins of all kinds, spars, and masts. After 1676, the coast of Maine east of Wells was abandoned by its besieged residents during the Indian wars, but such was not case west of Wells and along the southern New England coast. The booming West Indies trade was only one component of what was already a vigorous transatlantic trade. A closer look at the various items imported and exported from Boston in the century beginning in 1650 helps explain why the port of Boston was one of the most important in the English empire, and why the New England coast southwest of Wells was the center of shipping and trade and the very nucleus of the birth of a nation.

Dow (1935) provides a particularly detailed list of incoming and outgoing cargoes. In many cases, cargoes were imported from West Indies and colonial ports to the south, brought to Boston and then re-exported to Europe on outgoing merchant ships. In this context, one can deduce from Dow's descriptions of Boston's vigorous traffic in ingoing and outgoing ships that there were three main categories of trading goods: local products shipped out, imported products re-exported, and incoming merchandise, much of it not obtainable in the colonies (Table. 1). One interesting occupation, perhaps now long forgotten, was the active trade in tropical hardwoods, which were harvested by Yankee adventurers from the forests of the Yucatan Peninsula in the Bay of Campeachy. Dow (1935) notes a number of ships bringing provisions and supplies to New England woodsmen who stayed for months at a time in the Yucatan forests where they cut tropical woods that would then supply French, English, and Italian furniture-makers in Paris and other European cities, and later, English toolmakers in Birmingham and elsewhere. The net result of the work of the enterprising Boston merchants was to create a balance of trade that was so profitable for New England residents as to create an economy that grew by leaps and bounds for decades with only occasional interruptions. The contrast to United States trading deficits during the last few decades of the 20th century and the first seven years of the 21st century is a disturbing reflection of the rise and decline of a vigorous American industrial economy that had its roots in the booming shipbuilding communities of colonial New England. The litany of commodities exported, imported, or re-exported from Boston noted by Dow helps explain why Boston was America's most important port in the late 17th and early 18th centuries.

Table 2. Exports and Imports

Locally produced exports	Imported goods	Imported goods later exported
Naval stores	Brandy	Tobacco
Masts	Wine	Sugar
Timber	Cloth	Indigo
Codfish and Mackerel	Ironware	Cotton
Cattle	Bar iron	Wool
Provisions	Hand tools	Ginger
Beaver fur	Rum	Tropical log wood
Moose hides	Sugar	Fustic (and other dyes)
Deerskin	Indigo	Rum
Log wood	French commodities	Wine
Horses	Candlesticks	Linen
Beef	Tin lanterns	Salt
Pork	Nutmeg graters	Ironware (to southern colonies)
Butter	Rat traps	Cloth
Cheese	Woodenware	Pig silver
Flour	Spoons	Silver plate
Peas	Beer caps	Pieces of eight
Biscuits	Hand shears	Spices
Rum	Sucking bottles	Chocolates
Linsey-woolsey	Milk trays	Raisins
House frames	Lead shot	Oranges
Staves	Spices	Figs

Locally produced exports	Imported goods	Imported goods later exported
Salt boxes	Chocolates	Almonds
	Raisins	
	Oranges	
	Tar	
	Turpentine	

Of particular interest are the number of items such as salt, rum, raisins, spices, almonds, wine, wool, cotton, indigo, sugar, and tobacco, which were imported for local use and, at the same time, exported as trade items. Many a New England coasting captain made a small fortune shipping house frames, cotton, pork, fish, and staves, from which rum and molasses kegs were made, to the West Indies ports and then returning with molasses and sugar, which would then be made into rum in New England distilleries. Since rum was also made in many other locations, enterprising traders may have brought it with wine to Europe, then traded it for French brandy, which was consumed in significant quantities by both English and colonial merchants.

Enterprising New England captains also brought another commodity not mentioned by Dow (1935) from one location to another: slaves. Dow notes that "Negros," who worked as slaves in the sugar plantations of the Barbados, were the consumers of the lowest grade of codfish. The critical role of these slaves as facilitators of New England's lucrative trade in rum was not discussed until the last few decades of the 21st century. The sugar cane harvested with slave labor was only surpassed in importance by the neutral trade and the cod fishery in helping to build those stately clapboard houses on tree-lined New England streets.

Items such as tar and turpentine were imported from southern colonies and then incorporated in shipbuilding; tar was essential to the safe waterproofing of any wooden ship. These newly built ships were then either used for the coasting and transatlantic trade or were resold to English and other merchants as one of New England's most important domestically produced commodities. Other New England produced commodities of importance were lumber, spars, staves, and other woodenwares, which were shipped south to ports such as Charleston, SC, and especially to the West Indies. After 1765, Maine became an increasingly important source of these forest products.

While commodities trading patterns were complex and involved ports throughout the Caribbean and Atlantic Oceans, the most important exports were derived from New England's resource-based economy. Codfish were its single most valuable trade item, as illustrated by the huge number of small fishing craft working the Grand Banks off Cape Sable. Dow (1935) does not mention the rich inshore fisheries, which range from the Bay of Fundy along the Maine coast to Massachusetts Bay and then southward off the coast of Cape Cod. Codfish and mackerel not only provided New England families with a reliable source of food throughout long winters but also constituted the single most valuable commodity traded for large quantities of cotton, sugar, and molasses imported from southern ports. The second most valuable New England export were forest products, especially in the form of masts, spars, house frames, and the dozens of products of the New England cooper, of which barrel and keg staves were the most important. Underlying the success of this trading economy were the anonymous shipsmiths, who "ironed" the colonial merchant and fishing fleets, and the bog iron bloomeries of southeastern Massachusetts, which, for almost a century, made this region the iron-producing center of the colonies. The forges and blast furnaces of these smithies and foundries have disappeared with hardly a trace, except for occasional references in obscure town histories. One might argue that they are just a mythical invention were it not for the tools, hollowware, and cannon balls that have survived from the colonial era in New England barns, cellars, tool chests, and workshops.

English or Indigenous Edge Tools?

Many contend or imply that all hand tools were made in England before 1800 (Bolles 1878, Gaynor 1997, Gordon 1996), an assertion that is both myth and reality. The myth is perpetuated by the reality of centuries of dependence on English and German saw steel and English carving tools and plane blades, which didn't end until the mid-19th century. Centuries of output by skilled English and German toolmakers preceded the colonial shipsmiths and furnished many of the tools for the first colonial settlers. Nonetheless, colonists made large quantities of tools in New England for their own use. The mystery remains as to where and when the colonial shipsmiths and blacksmiths began to make edge tools, which, if not brought by individual immigrants, were difficult to order from England as needed. To what extent did indigenous tool production supplement imported English hand tools offered for sale in shops such as George Corwin's shop in Salem, MA, in 1651?

> This shop, a few years later, was supplying the town with such articles as combs, white haft knives, barbers' scissors, flour boxes, carving tools, carpenter's tools of all kinds, door latches, curry combs and brushes for horses, and a great variety of earthen and woodenware. (Dow 1935, 153)

There can be no argument that substantial quantities of English hand tools were imported from England throughout the colonial period and well into the 19th century. The wide variety of English-made tools surviving in New England tool chests, workshops, and collections attest to the importance of and respect for English-made tools in colonial America. But George Corwin's shop wasn't the only source of tools for Salem artisans in 1651. Only a few miles south of Salem, Joseph Jenks was already making tools at the blacksmith shop at the integrated ironworks at Hammersmith (The Saugus Ironworks). Other furnaces and forges, if short lived, had already been established at Furnace Brook in Quincy and possibly on the Monatiquot River in Braintree. The Saugus Ironworks certainly supplied pig iron, for fining into wrought and malleable iron, and the fined iron bar stock itself to numerous community blacksmiths and shipsmiths whose identities have been lost to us. The wide variety of primitive horticultural tools and ironware that survive from this era, the well documented presence of numerous experienced ironworkers after the great migration, and the compelling necessity of domestic production of essential horticultural implements and iron ship fittings attest to the rapid evolution of a robust toolmaking community of artisans in 17th century coastal New England. Volume 8 of the *Hand Tools in History* series (Brack 2008) contains a sketch of the relative scarcity or abundance of common 18th and early 19th century English and American hand tools that have survived to be salvaged in the 20th century during the author's decades of tool-salvaging activities in New England.

The exquisitely fashioned tools made in 18th and 19th century England for the English gentry described in Roberts (1976) and Smith ([1816] 1975) obscure the robust activity of colonial era toolmakers. The heavy-duty, special purpose adzes, slicks, hewing and broad axes, and mast shaves needed by colonial New England shipwrights may have lacked the finished look of England's finely made Sheffield cast steel tools, but they nonetheless served their intended purpose well enough to make New England the world center of wooden shipbuilding from the colonial period to the late 19th century, with New York, Pennsylvania, and the Chesapeake Bay important but distant runners-up.

The shipsmiths, who made these edge tools before 19th century trade specialization gradually eroded the key role they had played in New England's maritime economy, are now entirely forgotten. The gradual evolution of edge toolmaking as a specialized trade is also nearly undocumented. Both the shipsmith as edge toolmaker and specialty edge toolmakers signed many of the tools that they forged, especially after the 1800s; their signatures testify to their existence. We are still not able to document the subtle evolution of this trade specialization. Maritime histories as important as Baker (1966, 1973) and Morison (1921, 1930) hardly note their presence. *The Oxford Companion to Ships & the Sea* fails to include the words shipsmith or "ironing" in its definitions and mentions iron as something introduced to the shipbuilding trade with the coming of the Industrial Revolution, despite the fact that, in its description of shipbuilding in the Middle Ages, it refers to the bolts used in war ships to lock their timbers together (Kemp 1976). The diversity and unique history of the strategies for producing the steel that shipsmiths and edge toolmakers used to make edge tools that built the wooden ships of a new nation are almost as forgotten as the New England shipsmiths themselves. Sufficient information exists in a wide variety of sources for us to at least attempt a reconstruction of steel- and toolmaking strategies and techniques before and after 1607.

II. Ferrous Metallurgy Before and After 1607

Historical Background

> The year 1776 saw two great revolutions unfolding. In Europe the Industrial Revolution was beginning to change both the face of the earth and the lives of people everywhere. In America a political revolution was bringing forth a new continental empire. The revolutionary generation of Americans was born in a country which labored with a medieval technology in a colonial economy... (Pursell 1981, 1)

Iron-smelting in North America is rooted in a mixture of ancient European steel- and toolmaking strategies and techniques. These derived from ever more ancient and esoteric traditions in Mediterranean, Muslim, and Chinese cultures. The earliest iron forge in North America subject to archaeological excavation and carbon dating is the primitive Viking bloomery at L'Anse aux Meadows, Newfoundland (c. 1025). Viking shipsmiths used it for the production of iron fittings for repairs on their sailing vessels. Weapons production or repair was a likely secondary function. The forging of malleable iron horticultural tools was another possible use of this forge only if L'Anse aux Meadows was the location of a semi-permanent settlement where crop cultivation occurred. The direct process production of iron in this primitive bloomery bowl furnace was truly a medieval technology.

One of the first documented acts of shipbuilding in continental North America occurred at the Popham settlement (Fort St George) on the Kennebec River, in 1607 – 08. The tools used there to construct the pinnace *Virginia* were almost certainly made in England, at one or both of England's active ironworking centers. The two principal centers of smelting and tool production in the 16th and early 17th centuries were the Forest of Dean on the River Severn, with its associated upstream midland forges and furnaces, and the Weald of Sussex, south of London, which was the center of cannon production in Tudor England. The techniques used to manufacture tools and weapons made in Renaissance England, which were either forged near the furnaces that supplied the smelted iron and forged steel for their production or were forged at nearby urban centers such as London (Sussex) or Birmingham (Forest of Dean), were *not* based on medieval technologies. The Italian Renaissance in central Italy, the German Renaissance in southern Germany, and the Renaissance of northern France and the Netherlandish communities to the north were characterized by a growing awareness of multiple strategies for producing steel and tools, including the use of the blast furnace. These robust Renaissance economies demanded and were dependent on the ever increasing production of iron and steel for ordnance, hand guns, tools, hardware, and equipment such as printing presses, wagon axles, and tilt hammers for water mills. The blast furnaces that supplied this burgeoning demand for iron and steel were a post medieval development, the first of two stages of an early period of industrial innovation, which culminated in the classical Industrial Revolution of cast steel, steam engines, coke-fired reverbatory furnaces, and rolling mills after 1775.

The tools and iron hardware used to construct colonial Maine's first ship, the pinnace *Virginia*, were made by the shipsmiths and edge toolmakers of the River Severn and the Sussex Weald, utilizing the long established continental traditions of making "German" steel from fined (decarburized) cast iron. The construction and use of cementation furnaces to produce blister steel was a second step in this early stage of the Industrial Revolution, which occurred in England only a few decades after the construction of Maine's first ship. First described by continental sources in the late 16th century, the first known cementation furnace was constructed at Nuremburg in 1601 (Barraclough 1984a, 13). Ironically, the cementation furnace would never supercede the well established tradition of fining (decarburizing) cast iron to produce "German steel" on the European continent, where, in the southern German-Austrian area, the use of the cementation process never became the principal steelmaking strategy.

In England, in 1607, no cementation furnace had yet been constructed. In 1613 – 1617 this process was patented in England. Barraclough (1984a) provides a detailed description of the failed attempts of William Ellyott and Mathais Meysey to patent and then replicate the production of blister steel in the early 17th century, which Germany had already produced and would soon be made in Danzig and Sweden. More successful but not completely documented was Sir Basil Brooke's probable production of blister steel around 1635 in the Forest of Dean just to the north of the River Severn. Smelting the phosphorus-free deposits of the Forest of Dean that Henry Bessemer used in his first experiments two centuries later, Brooke may have made blister steel from charcoal fired pig iron produced at his furnaces near Linton. Brooke moved his ironworks further north to Coalbrookdale in 1638, located in the Midlands area north of the Forest of Dean that was soon to become England's major iron producing region, along with Newcastle on the far northeast coast of England, by the late 17th century. The first documented use of the cementation furnace to make blister steel occurred in this area at Stourbridge in 1686, after the English Civil Wars and the seizure of Sir Basil Brooke's furnaces, when Ambrose Crowley, (father of Sir Ambrose Crowley, who was to establish steelmaking in the northeast at Newcastle) was recorded as making blister steel (Barraclough 1984a).

The two advantages of producing steel in the cementation furnace compared to decarburizing cast iron in a finery furnace lie in the tonnage of steel produced in the larger cementation furnaces versus that in a finery furnace (2 – 5 tons vs. ½ ton), and the ability of the forge master to determine the carbon content of the steel being produced by calculating and controlling the time of the firing. Charcoal iron baked for three to five days had a radically different microstructure (carbon diffusion pattern) than bar stock fired from ten to twelve days. Under any circumstances, carbon diffusion derived from the layers of charcoal packed between the iron bar stock was gradual. The innermost layers of iron in the cementation furnace had the lowest carbon content, while the outer layers had the highest. Unlike cast steel, blister steel was never characterized by a homogenous (evenly distributed) carbon content.

Prior to 1607 and the Popham expedition, the continental tradition of steel production in England by fining cast iron accompanied the introduction of the blast furnace in the late 14th century, an event that occurred a century or more after its widespread use in Europe. It is possible that the nearly phosphorus-free ores of the Forest of Dean and the manganese laced siderite ores of the Sussex Weald postponed the need for blast furnace produced iron. Inefficient bloomery furnaces were smelting iron and raw natural steel in the Weald for centuries prior to the Roman conquest of Britain. The Roman era iron industry in southern England and the later robust Tudor era ironworking industry of the Weald were based on rich but limited natural resources that were depleted by the late 17th century. These resources may have been a principal motivation for the Roman occupation of Britain (Cleere 1985). The blast furnaces in the Sussex Weald (1496) and in the forest of Dean (possibly established in the mid-15th century) signal the first stage of a post-Medieval Industrial Revolution that was quickly transported to North American in the great migration to Massachusetts Bay in 1629 – 1643. When the shipwright Digby from London began building Maine's first ship at the mouth of the Kennebec River in 1607, he was utilizing tools made by steel and toolmaking strategies and techniques of a post-medieval Renaissance that had been well established for 150 years. When Hammersmith, the integrated ironworks now known as the Saugus Ironworks, was constructed to the north of Boston in the 1640s, it was as modern and up-to-date as any integrated ironworks in England or Europe. The experienced ironworkers and forge masters, including shipsmiths, who manned it were already residents of the Massachusetts Bay Colony and ready and willing participants in its construction and operation (Hartley 1957).

Many smaller direct process bloomery furnaces and anchor forges were soon established throughout the bog-iron-rich communities of southeastern Massachusetts. The function of the integrated ironworks at Saugus or of the smaller bloomery forges in early colonial America, some of which may remain forever undocumented, was the production of cast iron, hollowware, and especially wrought and malleable iron, either from fined pig iron or bloomery "muck bars" for two purposes. The production of the horticultural tools for a burgeoning agricultural economy was an absolute and unequivocal need; hoes, shovels, picks, forks, as well as iron pots, fire backs, hardware, and other tools were all essential for the survival of the first colonial settlements. The second function of the Saugus

Ironworks and the poorly documented direct process bloomery furnaces and forges, which supplied the shipsmiths in coastal communities, was the equally essential, but less obvious, necessity of producing the iron fittings for a blossoming colonial shipbuilding industry. This industry grew out of the inescapable need to establish a viable fishing industry and inter-colonial coasting, West Indies, and transatlantic trading routes. These mercantile endeavors were based upon the rich natural resources of forest and fisheries, which could only be exploited and transported in the context of a mobile maritime community. The efficient functioning of this resource-based economy was, in turn, dependent on the ironware and edge tools produced by the bloomsmiths (iron smelters) and shipsmiths who gradually spread through the coastal plains of the American colonies in the 18th century. The critical trades that determined the success or failure of New England's maritime endeavors and its shipbuilding industry were thus those of smelter, shipsmith, and shipwright. If we follow the dots, we can trace New England's robust trading economy back not only to colonial era shipwrights, edge toolmakers, and shipsmiths, but to an odd geological fluke: the legacy of the glacier-derived bog iron deposits of southeastern Massachusetts. These indigenous iron ore deposits played as important a role in the late 17th century New England economy as imported English and European iron, steel, and hand tools.

Moxon's ([1703] 1989) observations summarize late Renaissance sentiments about the essential role of ferrous metallurgy in a rapidly expanding pyrotechnic society.

> *For the reason aforesaid I intend to begin with* Smithing, *which comprehends not only the* Black-Smith's Trade, *but takes in all* Trades *which use either* Forge *or* File, *from the* Anchor-Smith, *to the* Watch-Maker ; *they all working by the same* Rules, *tho' not with equal exactness, and all using the same tools…* (Moxon [1703] 1989, preface 6)

The sequential exploitation of colonial America's rich iron deposits facilitated the success of the American Revolution by providing iron, the essential raw material for gunsmiths, cannon and cannon ball founders, for edge tool production for timber harvesting and shipbuilding and the forged iron hardware for New England's colonial shipbuilding industry. The obscure role of the shipsmiths may at first seem inconsequential, but the ships built with their assistance were the key to the success of the colonial economy. This sequential exploitation of America's rich iron deposits began with the bog iron deposits of southeastern Massachusetts and then extended to the Pine Barrens of New Jersey and to the rock ore deposits of western Connecticut, Pennsylvania, Maryland, and New York.

The principal unanswered questions pertaining to the rise of a vigorous indigenous colonial shipbuilding industry are: to what extent and where did the New England shipsmiths of the late 17th and early 18th centuries begin producing edge tools for American shipwrights? Where did they obtain the steel used to forge weld their edge tools? Traditionally it has been assumed that these tools were made in Sheffield and imported to America. But, the great majority of surviving heavy duty shipwright's tools recovered from New England tool chests – adzes, hewing axes, mast shaves, slicks, and timber-framing tools – appear to be American-made. In contrast, almost all plane blades, carving tools, and smaller woodcutting tools from this era appear to be English, or in some cases, German. Tools made in England are almost always marked with the imprints of their maker. The middle range of edge tools such as spoke shaves, draw knives, socket chisels, and gouges recovered from New England tool chests appear to be a mixture of English Sheffield-made edge tools, and more primitive American-made specimens. For every finely made, signed English draw knife, half a dozen, often unsigned, weld-steel draw knives make their appearance in tool kits dating from the two centuries before Witherby, Kimball, and Wilkinson began mass production of their finely made cast steel drawshaves.

The great migration to the Massachusetts Bay Colony and other New England ports ended quickly with the onset of the English Civil War (1642-1648), the turmoil of the last years of the House of Stuart and the establishment of the Commonwealth (1649 – 53), and Cromwell's protectorate (1653 – 1659). The struggle for the control of England between the monarchy and Parliament, which was the cause of the Civil War, also ended colonial access to the English shipping industry. The roots of a vigorous colonial shipbuilding industry arose in response to the urgent need to replace English ships in the incipient coastal, West Indies, and transatlantic trades. New England's economy was based on the triad of forest products and timber harvesting, fish and fishing, and ships and shipbuilding; the economic lynchpins of an economy that, without sea transportation, would have no markets beyond the survival needs of the newly arrived immigrants. A review, or at least a sketch, of Contact Period ferrous metallurgy is an essential component of a reconsideration of the importance of the New England shipsmith in the colonial economy, in so far as four hundred years of separation allows us to attempt this review.

When John Winthrop Jr. and Robert Bridges went to England (1641 – 1643) to petition the court of Charles I to allow construction and operation of the ironworks at Saugus, organized as the "Company of Undertakers for the Ironwork" (Bining 1933), they were among many participants in the rapidly expanding market economy of a Renaissance when exploration and settlement of the New World was totally dependent on innovative and efficient *post-medieval* strategies and techniques for producing iron and steel weapons and hand tools. The rise of the shipsmith first in continental Europe and England and in colonial America after 1640 was an essential component of the growing industry of ferrous metallurgy, the production of tools and weapons from smelted iron ore by the mechanical and thermal treatment of wrought iron, malleable iron, raw steel, and cast iron. The manufacture of iron hardware and edge tools for the construction of colonial era ships was the critical first step in the successful organization of the coasting West Indies and transatlantic trades, which were essential for the survival of the newly established colonies. A more detailed survey of the diversity of steelmaking strategies of the historical milieu in which Maine's first ship, the pinnace *Virginia*, an icon of icons, was built, provides insight into both the origins and traditions of the New England shipsmith, forefather of all the famed edge toolmakers of the Gulf of Maine and southern New England's toolmaking centers.

American Beginnings: A Diversity of Steelmaking Technologies

European shipsmiths labored for centuries in obscure river and sea-side workshops, forging edge tools and iron fittings, which made exploration and settlement of the new found lands possible. Migrating to New England and elsewhere, those now forgotten shipsmiths were the hidden enablers of the robust colonial maritime trading economy, which evolved into the world's most powerful industrial nation in the late 19th century.

All empires rise and fall: In the post-World War II era, enterprising far eastern countries began the successful imitation of the American factory system. English tool manufacturing was in rapid decline by the early 20th century, and the decline of American toolmaking soon followed. By the early 21st century, third world countries began to produce hand tools far more cheaply than the United States could. The cycle that began at Halstadt (Austria) with the accidental production of high quality iron and then steel tools from the spathic ores of the Erzberg (ore mountain) in the early Iron Age moved from the hegemony of one culture to the next: Rome (natural steel from Noricum), Germany (German steel), England (cementation, crucible, and alloy steels), America (crucible, alloy, and bulk process steel). Japan, Taiwan, and China now dominate world hand tool production, making articles of lesser quality but cheaper in price.

In the modern era of hundreds of sophisticated alloy and tool steels, only a handful of small companies in Japan, Sweden, Switzerland, and the United States make high quality woodworking hand tools. In a post industrial society, armed with sophisticated new methods of information retrieval and communication such as the World Wide Web, can industrial toolmakers master the art of the edge tool? Is the work of the shipsmith and edge toolmaker worth remembering, rediscovering, documenting, and consecrating, in view of the essential role they played in our political and industrial independence? Most importantly, will a viable market for edge toolmakers re-emerge in the post-industrial era of the creative economy, especially in locations such as Maine, with its undying tradition of wooden boat and ship building, furniture making, and post and beam timber-framing construction?

Ancient strategies for making steel edge tools don't succeed each other in neat sequential patterns; they overlap with messy variations that even expert archaeometallurgists have difficulty in interpreting. The same observation can be made about the "myth of the Industrial Revolution." Despite the virtual tsunami of industrial change in America between 1837 and 1850, the Industrial Revolution in both England and America is the story of a long series of gradual, incremental, innovations and inventions over a period of centuries. Ancient steel- and toolmaking strategies and techniques lingered and overlapped with more modern techniques and innovations. Our knowledge of early steelmaking techniques during the early years of the gradual evolution of modern industrial society is more protohistoric than scientific. We can study the microstructure a few surviving steel tools. Much of our knowledge of early techniques is from lore, story, or observations of early forms not available for microstructural analysis. In the form of the lingering accidental durable remnants from an early age, edge tools tell a complicated story that contradicts all linear models of technological change.

When Champlain sailed for New France in 1604, he reached the new found lands in less than four weeks. Few explorers left such detailed descriptions of their explorations as he provided in his journals. Yet we forget to ask what tools were used to build his ships or those of the Elizabethan adventurers who preceded and followed him. By 1600, early modern steel manufacturing technologies had appeared that supplanted ancient methods of forging. The world wide exploration and empire building of Spain, the Netherlands, France, and England would not have been possible without the bulk production of cast iron, which could be decarburized to make steel or refined into wrought iron and then carburized into steel weapons and edge tools. Knowledge of the ferrous metallurgy of the tools so essential for the settlement of the New World is now a nearly forgotten chapter in this history, yet ancient steelmaking strategies survive into modern times, nearly invisible in the glowing industrial landscape of our multiple Industrial Revolutions. Four different strategies characterize steelmaking efforts at the time of the exploration of the new found lands, the construction of Maine's first ship the pinnace *Virginia*, and the settlement of North America. These strategies characterized all edge tool production prior to 1600. The four basic techniques used to make steel in 1600 were forging, carburizing, decarburizing, and fusion. They are best summarized by Wertime (1962) as three fundamental strategies of making steel: adding carbon (forging and carburizing), subtracting carbon (fining cast iron), or mixing or fusing the carbon content of wrought and cast iron, as in Brescian steel.

The principal method of steel production in the Renaissance was the decarburization of cast iron, a relatively new strategy that accompanied the coming of the blast furnace, soon supplemented by the cementation furnace, which allowed the bulk production of blister steel by the carburization of multiple bars of wrought iron. Benjamin Huntsman's clever adaptation of the obscure art of crucible steel production in 1742 was still far in the future. While the English production of both blister and crucible steel is a well known chapter in our industrial history, the continental strategy for producing steel by decarburizing cast iron (German steel) is nearly forgotten, especially in English speaking countries. An extensive collection of axes and other edge tools, almost all of which are made from German steel, can still be seen today at the Maison de l'Outil museum in Troyes, France, illustrating the widespread use of the continental method of making tools with German steel before 1900. Before the appearance of the blast furnace in 1350 and the widespread production of German steel from decarburized cast iron lies an even more obscure chapter in the history of ferrous metallurgy, the root of all steel production strategies in the ancient tradition of smelting natural steel.

Natural Forged Steel

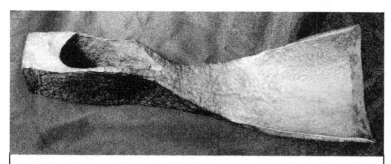

Figure 24 Unusual forge welded natural steel mortising adz, use unknown, circa 1700, 10 1/4" long, 3 ½" wide mouth, 1 ¼" deep curved blade. In the collection of the Davistown Museum 71903T1.

Among the most ancient strategies for making steel is that of producing natural steel from a direct process bloomery furnace. Blast furnace production of iron evolved after 1300, in part due to the inherent wastefulness of the direct process bloomery, where 50 to 65% of the iron in the ore being reduced was lost in the smelting process. Only small kilogram quantities of iron or raw steel could be produced as the loupe derived from a direct process bloomery. However, direct process iron production was simple and could be easily executed in any isolated location with access both to wood or charcoal fuel and iron ore. For centuries, desert tribesmen and isolated central European bloomers made iron in simple bowl and shaft furnaces, hammered it into sheets of iron, and slowly carburized it into sheet steel under layers of charcoal that would protect it from the deleterious impact of oxidizing gasses. Any blacksmith could increase the fuel to ore ratio of his bloomery furnace, slightly altering the reduction of the iron ore to wrought iron by increasing the amount of carbon available from the burning fuel. The addition or removal of the carbon from iron ore during smelting was a gradual process and was dependent on both the amount of charcoal used in the firing and the temperature of the firing. High temperatures resulted in the absorption of carbon by iron, creating unwanted cast iron with as much as a 4% carbon content. As with fined cast iron, knowledgeable bloomsmiths could selectively remove enough carbon to make iron that would qualify as raw steel (0.05 – 1.0 % cc) or change furnace conditions to encourage the removal of most charcoal-derived CO (carbon monoxide), producing the most common products of a direct process bloomery, wrought iron (0.02 – 0.08% cc) or malleable iron (0.08 – 0.2% cc). Iron produced in the bloomery with a carbon content greater than 0.2% qualifies as natural steel and differs from modern "low-carbon steel" only in its high siliceous slag content.

It is interesting to note the widespread production of natural steel throughout southeast Africa by Bantu immigrants or their descendents in very simple bowl furnaces as early as the first millennium (Van der Merwe 1980). We don't normally think of Africa as a steel-producing region at this time, but the fact that raw steel was produced in this location at this early date helps illustrate the adaptability of the bloomery process and its potential to be manipulated to suit the metallurgical needs of the communities which it served. Van der Merwe (1980) has this comment on African iron-smelting:

> Its evolution followed ideas which were highly observational and inventive, producing a wide variety of approaches, furnace designs, and smelting products… African iron-smelting cannot be described in terms of either the direct or indirect process because its primary product is neither wrought iron nor cast iron. Instead, a bloom of *high carbon steel* was produced directly from the smelting furnace and subsequently *decarburized* in the forge. (Van der Merwe 1980, 486)

Tylecote (1987) also illustrates the adaptability of the bloomery furnace:

> Two grades of iron can be produced in the Catalan hearth: soft iron or natural steel. The latter is made, as expected, with a higher fuel/ore ratio and slag tapping is more frequent so that the Fe is not decarburized too much in the slag bath. More time is needed and manganese must be present in the ore. This is in agreement with the ideas on Noric iron, but in neither case does the Mn appear in the iron; it merely alters the equilibrium so that a higher carbon iron is in equilibrium with a lower iron slag which is made fluid with MnO rather than FeO.

> …reduction takes place to high-carbon iron (cast iron mixed with slag). This iron is oxidized lower down and finishes in the solid state while the fluid manganiferous slag liquates to the bottom. Frequent slag tapping will avoid over-oxidation of the metal and allow carbon to be retained in the metal, as desired.

> The blast from the trompe is saturated and no doubt the water vapour takes part in a water–gas reaction:

> $$H_2O + C \rightarrow CO + H_2$$

> The hydrogen formed accentuated the reducing reaction in the higher levels and produced a more steely iron. It has been found that the use of blowing cylinders instead of the trompe results in a more ductile iron. (Tylecote 1987, 167)

Iron currency bars, the source of the iron bar stock used by blacksmiths and sword and edge toolmakers and traded over long distances, and tools recovered from archaeological sites in central Europe testify to the fact that making raw steel in a direct process bloomery furnace dates from the first appearance of the Iron Age at Halstadt in central Europe. Agricultural tools recovered from central European sites and dating to pre-Roman times and the currency bars shipped to many areas of central Europe from the iron-smelting centers of Austria often contained surprisingly high levels of carbon, i.e. enough to qualify them as natural steel artifacts (Pleiner, 1980, Wertime, 1980). Though not homogenously distributed, the carbon content of these tools and currency bars means that they cannot be defined as wrought iron, i.e. containing less than 0.08% carbon.

56

Making natural steel from a direct process bloomery in the Forest of Dean or the Weald in Sussex would have been one of three well known options for steelmaking at the time of the construction of the pinnace *Virginia*. Few, if any, written sources have survived to tell us which techniques Bristol and south coast England blacksmiths used to fashion the edge tools used to build the *Virginia*. At least some of the edge tools made at this time could have been made from natural steel, even though German steel was the dominant steelmaking strategy.

Figure 25 Socket chisel, natural steel, 9 ¼" long, 15/16" wide, with a 3 ¾" long replaced handle. In the collection of the Davistown Museum 41907T1.

A West Virginia bloomer making an edge tool at an isolated forge in 1865 had something in common with a Hittite blacksmith making a sword for battle in 1200 BC. Both might have searched for nodules of a mysterious shiny hard substance entrained in the soft bloom of wrought or malleable iron in their primitive bowl or shaft furnaces. Or they may have altered their fuel to ore ratio, allowing a heterogeneous bloom of raw steel to form, instead of low carbon malleable or wrought iron. By extracting these nodules of natural steel or reworking this bloom of raw steel, being careful not to "burn" the steel by excessive heating and oxidation, both smiths could fashion small bars of raw steel by repeated hammering and reheating, expelling unwanted slag and producing, at least in the case of the early Iron Age smelter, the ubiquitous currency bars of iron and raw steel that were traded throughout Europe in the early Iron Age. It was the experienced smith, swordsmith, shipsmith, or edge toolmaker, who knew the art of tempering and quenching, that most important step in converting raw natural steel into functional, durable tools and weapons. Smith's could use this steel bar stock to make an all-steel edge tool (rare) or, more commonly, forge it as the steel edge of a serviceable chisel or sword. These tools were composed of natural steel, hand forged by our mythic blacksmiths and sword-makers into tools or weapons we might see in a museum.

A Roman blacksmith at Noricum, the Styrian province of Austria and major steel-producing center of the Roman Empire, might have had lucky access to the siderite iron ore from the Erzberg (Ore Mountain). This ore, high in manganese, neutralized unwanted sulfur as slag at a lower smelting temperature and thus helped forge masters control the reducing atmosphere of the furnace interior. The result would not be iron with entrained nodules of steel, but the more systematic smelting of raw steel blooms with heterogeneous carbon content. So too, bog iron bloomers from Taunton, Bridgewater, or Raynham might have had access to bog iron with a modest manganese content, facilitating their discovery of steel nodules in the siliceous bloom of bog iron. In most cases, toolmakers would use nodules or blooms of natural steel for the cutting edge of woodworking tools or they would hammer and reforge heterogeneous blooms of raw steel into serviceable edge tools.

Figure 26 Broad ax, 17th or early 18th century, forged iron, 9 ½" long, 8" wide blade, found at Alna, Maine. Now in the collection of the Davistown Museum 111001T29.

There are no written records of tool production in prehistory, but lingering stray tools, such as a chisel at the British Museum circa 100 BC, attest to the fact that natural steel tools were forged almost as soon as the first blacksmith hammered a bloom of wrought iron. It didn't take early ferrous metallurgists long to realize that their primitive steel swords made of pattern welded raw steel and iron were superior to the soft malleable iron swords that bent so easily in battle. In North America, the direct process bloomery production of wrought iron lingered in rural areas into the early 20th century. Implicit in the presence of these bloomeries is the smiths' knowledge of the ancient craft-based tradition that, if one ran a bloomery furnace at a hot temperature with a higher proportion of charcoal to fuel, the carburizing effect of burning charcoal could produce nodules of natural steel in the bloom of iron, a bloom of mostly raw steel, or more often, durable malleable iron with the same carbon content as modern "low carbon steel." Streams of liquid cast iron, which would solidify outside the furnace into useless lumps of solid cast iron, were sometimes an unwanted byproduct. Forge masters and blacksmiths who made iron and steel tools from the products of Europe's early furnaces or America's numerous bloomeries left no written records behind to tell us of their knowledge of ancient forging techniques or to document the small quantities of hand tools they produced for the local market economy, yet the tools they made still survive, often unrecognized, the rusty relics at the bottom of old tool chests. High in siliceous slag content, the forge welded raw steel or malleable iron tools they made often have the same carbon content (0.08 – 0.5% cc) as modern low carbon steel and, occasionally, the higher carbon content of modern tool steel (> 0.5% cc).

Steel made from Carburized Iron

The 256 years between 1607 and the 1863 Sanderson price list (Fig. 27) was a period of overlapping steelmaking strategies. While German steel production initially dominated English and European production before 1700, the cementation process dominated English steel production techniques after 1700. Both produced steel that was welded onto the cutting edges of woodworking tools that were first made in England and then in colonial New England and then used to build colonial America's great fleets of fishing, coasting, and merchant vessels. But colonial woodworking toolmakers still used the earlier techniques of forging natural steel, carburizing forged iron, and case hardening to produce woodworking tools.

One of the most ancient of steelmaking techniques was hammering a bloom of wrought iron into thin sheets, followed by emplacing it in the coals of a bowl forge or even a desert campfire, taking care to avoid contact with combustion gasses. In some cases, clay or other materials covered portions of the iron not to be carburized. Carburization, i.e. the addition of carbon to low carbon wrought or sheet iron to make steel, was a very slow process. Carbon penetration occurs at a rate of only a few millimeters every 8 hours. The resultant sheet of iron had a hard thin layer of steel on its surface with softer, low carbon iron under the thin outer surface of steel. Ancient forge masters or Wild West mountain men could pile, bundle, and pattern weld by hammering and reforging this material into primitive swords, knives, and other tools, but they could only make, at most, a few tools at a time. A second common variation in the case hardening process involved forging the bloomery iron into thin bar stock and submerging the bar stock in a charcoal pit fire, often for days. The result was raw steel bar stock that was further refined by hammering, piling, and reforging, producing a high grade of steel bar stock that was then forge welded as the steel edge on knives, chisels, scythes, and hammer heads.

The appearance of blast furnaces after 1350 marked the beginning of the post-medieval era of the bulk production of large quantities of cast iron. In late medieval Europe, when large blast furnaces began producing ingots of brittle high carbon cast iron, finers decarburized the ingots in their finery forges and then made bars of wrought iron (< 0.08% cc) or malleable iron (0.08 – 0.2% cc) or low-carbon steel (> 0.2% cc) in their chaferies. The blacksmith's task was to reforge this iron bar stock into useful tools. Implicit in the traditional and long-established method of fining and thus decarburizing cast iron to make wrought and malleable iron was the need to expand the tedious production of steeled edge tools made by hand forging or case hardening by the smelting of larger quantities of steel. Halting the decarburization of cast iron to produce steel was a strategy as old as the blast furnace itself (see below). But a new method for producing steel from the fined (decarburized) cast iron of the blast furnace was about to appear. After cast iron was sufficiently decarburized to produce malleable and/or wrought iron bar stock, it was the newly invented cementation furnace that was used to *recarburize* cast iron-derived wrought iron into blister steel. In 1607, cementation steel furnaces in England and then in the American colonies had not yet appeared. The need for a more efficient steel furnace for carburizing wrought and malleable iron was closely connected to the exploration, conquest, and settlement of the New World and to the shipbuilding, edge toolmaking, and gun founding industries that made colonial empires possible. The construction of the *Virginia* occurred just prior to the innovations in the production of steel from carburized wrought and malleable iron.

24　　　　　　　　*Sheffield steel and America*

Table 1.2 *Sheffield and Pittsburgh crucible steel prices 1863*

	Cents per lb		
	Sanderson Bros New York	Singer, Nimick Sheffield Works	Jones, Boyd Pittsburgh Works
Best cast steel	22	21	20–21
Extra cast steel	23		
Round Machinery	14	13–15	13–16
Swage cast steel	25		
Best double shear steel	22		
Best single shear steel	19		
Blister first quality	17½		
Blister second quality	15½	} 8–12	
Blister third quality	12½		
German steel best	15½		
German steel Eagle	12½	} 9–12	} 9–11
German steel third quality	11½		
Sheet cast steel 1st quality	22		
Sheet cast steel 2nd quality	18	} 15–21	} 15–23
Sheet cast steel 3rd quality	16		
Shovel steel best	14½		
Shovel steel common	13½		
Sheet cast steel for hoes	14½	11½	11½
Mill saw steel	15½	14	14
Billet web steel	17½		
Cross-cut saw steel	17½	18	18
Best cast steel for circulars to 46 in.	25	23	23
Toe corking best	10	9¾	9¾
Spring steel best	11		
Spring steel 2nd quality	10	} 9–10¾	} 9–10¾
Spring steel 3rd quality	8¼		

Source: SCL Marsh Bros. 249/24, 28–9.

Figure 27 Geoffrey Tweedale, 1983, *Sheffield Steel and America: Aspects of the Atlantic Migration of Special Steelmaking Technology, 1850-1930*. F. Cass & Co., London, England, Table 1.2, pg. 24.

In the late 16th century, German forge masters designed and built closed box converting furnaces that carburized wrought and malleable iron bar stock into blister steel. After a week of baking and being protected from combustion gasses, steel bars with higher carbon content in their outer layers and less steely inner layers resulted. Gasses (CO) that were formed in the converting furnace caused blisters on the surface of the steel, hence the name blister steel. For 150 years (1700 - 1850), the converting or cementation furnace was a principal method of manufacturing large quantities of relatively low quality steel, first in England, then in America. The high prices for blister steel on the Sanderson price list (Fig. 27) illustrate the energy and labor intensive cost of blister steel production. In colonial America and the early Republic, now forgotten converting furnaces were a component of a few forge sites and produced an undocumented percentage of the steel that was consumed in America beginning in the second decade of the 18th century. Blister and German steel imported from England and Europe continued to be the principal source of steel used in the colonies after 1720. There is no documentation of the quantities of blister steel produced in America before 1800, nor of the amounts of German and English blister steel imported to America from England and Europe. The welded steel edge tools produced in America at this time are ubiquitous in New England tool chests, shop lots, and cellars; the origins of their steel cutting edges will remain forever unknown, whether natural, Brescian, blister, or German steel.

Blister steel was reworked by piling, bundling, and reforging to produce high quality shear and double shear steel (also often called "sheaf" and, if rolled, "spring" steel) for special purpose applications. Either imported from England or made in colonial steel furnaces and then reforged into shear steel, this blister steel was the probable source of the steel used by New England's edge toolmakers to forge, for example, the ubiquitous New England pattern broad ax, which was an essential tool in 18th century shipbuilding. The same was true of heavy duty timber-framing tools such as corner chisels, slicks, and mortising axes.

After 1750, cast steel could be produced chemically by English steelmakers in their Stourbridge clay crucibles. However, cast steel was extremely expensive to produce and was much more practical to use for watch springs, razors, carving tools, and plane blades than for large unwieldy broad axes, slicks, and mortising tools. American toolmakers did not make cast steel until after 1860 when the problem of the domestic production of high temperature resistant crucibles was resolved by the invention of the plumbago (lead) crucible, but they sometimes made edge tools of equal quality to any of those made from cast steel by the artful reforging of blister steel into shear steel. The puzzle remains as to where 18th century colonial toolmakers obtained their steel.

The Sanderson (1863) price list (Fig. 27), which is for imported Sheffield and other steels, reflects the wide variety of special purpose steels available to blacksmiths, saw makers, shipsmiths, and toolmakers in the era prior to bulk process and alloy steel production. The steels listed in the Sanderson column of the Sheffield steel price list of 1863 weren't suddenly invented in 1858. This listing reflects the evolution of a variety of steelmaking strategies over a period of centuries. Many of the steels in the Sanderson list were already being produced at forges and furnaces at many locations in America. The Sanderson list of imported steels is a summary of the cumulative knowledge derived from hundreds of years of smelting and forging by thousands of English and northern European blacksmiths and toolmakers working since the appearance of the blast furnace in Europe in 1350. The result of this cumulative knowledge of ferrous metallurgy was the ability to forge steel with a wide range of carbon content. Differences in alloy content and heat treatment techniques (quenching and tempering sequences) combined with these variations in carbon content, encouraged the production of steels with subtle differences in microstructure, each suitable for a different function. It is most surprising to note that, at the time of the publication of this listing of steel types, an understanding of the chemistry underlying the variations in the microstructure, and, thus, the physical characteristics of steel, was still several decades in the future.

All the types of steel produced by carburizing wrought or malleable iron were based on the principal of adding carbon to iron in carefully controlled, enclosed environments and avoiding contact with combustion gasses. Shear steel production by reforging blister steel was particularly difficult because combustion gasses could easily decarburize the steel being refined. Knowledge of these techniques, but not of the chemistry that explained why they worked, was the critical component of the craft traditions of the forge masters, shipsmiths, and blacksmiths working both in Europe and America before the role of carbon in the formation of steel was well known.

The disorderly history of ferrous metallurgy leaves many unanswered questions. The identity and production location of many edge toolmakers remains unknown. Only the tools survive to let us know that history is more complicated than the modern myths about early steelmaking strategies might allow. This phenomenon is best illustrated by the irony of the simultaneous existence of two principal steelmaking traditions. The carburization of wrought and malleable iron to make blister steel faced stiff competition from the older tradition of making steel by decarburizing cast iron.

Decarburization

Decarburization of cast iron to produce steel is an ancient process that is documented in use in China during the first millennium BC (Needham 1958). With the appearance of blast furnaces in Europe producing large quantities of cast iron, the strategy of making steel from cast iron made its first continental appearance. Decarburizing cast iron is now a nearly forgotten technique of producing steel, in contrast to the recarburization of wrought and malleable iron during blister steel production. The older process is forgotten in the sense that the Bessemer process involves the rapid decarburization of huge quantities of liquid cast iron in spectacular bursts of combustion, which quickly burn out both carbon, silicon, and other contaminants. The development of the Bessemer process and the Siemens-Martin open-hearth furnace in the mid-19th century soon replaced not only the tedious production of blister steel and its daughter products, spring and shear steel, but also the long continental tradition of producing "German" steel from cast iron. The decarburizing of cast iron produced much of the steel used in Renaissance Europe before the use of the converting furnace for blister steel production became widespread after 1680 in England and in the American colonies after 1713 (Bining 1933). We can trace the roots of the tradition of decarburizing cast iron to produce steel back to China, although there may be no link between the early Chinese use of decarburized cast iron and its later use in Europe by the 15th century. Needham (1958) has written a detailed history of the Chinese steel industry, which was established at or before 700 BC. His documentation of a well-established tradition of malleableizing cast iron to produce steel agricultural implements is particularly significant. A sophisticated white cast iron adz with a decarburized steel cutting edge survives from this era, circa 500 BC (Barraclough 1984a, 30). Later knowledge of the technique of malleableizing cast iron, extensively detailed by the French metallurgist de Réaumur (1722), was known to European ironmongers and founders by the 17th century.

German steel production has its roots in the early Iron Age and in the fortuitous circumstance of Celtic smelters in Carinthia and Styria (now Austria) having access to manganese-laced "spathic" iron ores, which helped to expel deleterious sulfur in the slag of both the mountainside bloomery furnaces of the early Iron Age at Halstadt and La Téne and the later high shaft Stuckofen furnaces of the Hapsburg Empire. The manganese in the spathic ores served as a flux, absorbing sulfur and other contaminants at lower furnace temperatures and encouraged uptake and more homogeneous distribution of carbon during this smelting process. When blast furnaces began producing cast iron in the German tool-producing centers of Augsburg, Nuremberg, and Linz (Austria), the siderite ores from Styria, from which this cast iron was derived, were low in sulfur and high in manganese, resulting in the production of a form of cast iron called spiegeleisen. German forge masters were able to partially decarburize this manganese-laced cast iron into "German steel" in fineries and in puddling furnaces as soon as high shaft Stuckofen furnaces, which produced natural or raw steel for centuries, were enlarged to produce cast iron (> 1350). The development of fining cast iron to produce steel instead of making steel in direct process shaft furnaces was the gradual evolution of a steelmaking strategy driven by warfare and increasing demand, and was facilitated by the growth of furnace capacity. Not containing significant amounts of sulfur or having the sulfur expelled in the manganese slag, German steel was superior to most forms of natural steel and was cheaper to produce than cementation steel, which became its competitor in the late 17th century. Barraclough (1984a) provides the following description of the technique of smelting German steel, which was also the principal steel-producing strategy in France:

> The burning out of the carbon from the metal would follow, but the high manganese slag was less violent in its attack than the normal slag rich in iron oxide. Consequently the burning-out process could be more readily controlled and the removal of carbon could be brought to a halt at the desired stage. ...The operator in the sixteenth century had to go by his own experience, the appearance of the slag: its color and fluidity, and the nature of the metal: whether it was liquid, pasty or solid. ...The product if all had gone well would be a somewhat heterogeneous mass containing higher-carbon and lower-carbon areas, but capable of being hardened by quenching and of taking a good cutting edge. ...The product was sorted into three groups: true steel and the two less prized categories, known in the later French practice of the same type as 'ferdoux' (wrought iron) and 'ferfort', an intermediate 'steely iron', which hardened somewhat on quenching and had some value in the making of hoes, plowshares and other agricultural items. (Barraclough 1984a, 27)

As with natural, and later, blister steel, German steel could be mechanically and thermally treated to create a wide variety of special purpose steels for Augsburg watchmakers or Nuremberg armorers. The technique of producing steel by decarburizing cast iron was used throughout Europe, including Spain, Italy, Russia, and northern Europe. Steel production using the English process of blister and then crucible steel production only surpassed the tonnages of German steel in the mid-19th century when Sheffield, England, became the world's largest steel- producing center.

Day and Tylecote (1991) in the *Industrial Revolution in Metals*, provide data on the extent to which German steel, which they call natural steel in the following table, was still being manufactured in different countries in 1840. It is important to note that in Great Britain steel was no longer being produced from the decarburization of cast iron. In France, blister steel and German steel production was about equal, and, in the Austrian monarchy and the German Customs Union the decarburization of cast iron dominated steel production in 1840. It is unfortunate that this widespread strategy for producing steel is still called "natural steel" in this and several other texts. The decarburization of cast iron is, in fact, a radically different procedure from the production of raw natural steel by the bloomsmith in direct process shaft and bowl furnaces.

When Henry Cort redesigned the puddling furnace in 1784, he created an improved, more efficient method for decarburizing cast iron into large quantities of high quality wrought and malleable iron, which was then rolled and shaped for its intended use. The puddling, or reverbatory furnace, also provided an opportunity to create steel by decarburizing cast iron; this, in fact, was his original objective. Unfortunately, Cort and English forge masters were initially unsuccessful in using the puddling furnace to produce steel. Not knowing about the chemistry of steel production and ignorant of the role of manganese in helping control the intensity of the oxidation of the cast iron, English forge masters had little initial success in the controlled production of steel in the environment of a puddling furnace in comparison to the long-established and successful tradition of decarburizing cast iron to produce steel in finery furnaces that had been used in Austria and Germany for centuries. Barraclough (1984a) and others note that, by 1835, the use of the reverbatory puddling furnace had been implemented, first in Germany and then England, to produce steel by partially decarburizing cast iron. Between 1835 and 1870, successful use of the puddling furnace to make steel helped meet the rapidly increasing demands for steel in England, Europe, and America.

Producing Country	Natural Steel	Cementation Steel	Total Product
Great Britain	nil	20,200	20,200
Austrian Monarchy	12,600	nil	12,600
German Customs Union	7,000	100	7,100
France	3,320	3,710	7,030
Russia	520	2,640	3,160
Sweden and Norway	1,970	890	2,860
Spain	200	100	300
Italian Peninsula	100	100	200
TOTAL	25,710	27,740	53,450

300

Figure 28 Joan Day and R. F. Tylecote, eds, 1991, *The Industrial Revolution in Metals*. Brookfield, VT: The Institute of Metals, pg 300.

Documentation of the use of the puddling furnace, also known as an "air furnace," to produce steel in America is lacking, but the puddling furnace was a ubiquitous component in foundry complexes throughout the United States by the time of the Civil War. Its efficient production of wrought and malleable iron is well known. One of the more interesting variations in the use of the puddling furnace was the A.M. Byers Company's perfection and production of high quality wrought iron in Pittsburg in the early 20th century. The Byers Company used Bessemer steel and mixed it in the puddling furnace with carefully measured amounts of siliceous slag to produce its famed +/- 700 kg loops of wrought iron, a process graphically illustrated in its publication *Wrought Iron* (Aston 1939). Individual founders and forge masters could also have used the puddling furnace to make undocumented quantities of steel during their fining of cast iron. This process would, in part, explain the continual anomalous appearance of edge tools in New England tool chests not marked "cast steel," not drop-forged, and not obviously of a primitive bog iron or natural steel constituency, the latter usually characterized by a high silicon slag content. The demand for steel grew rapidly in the early decades of the 19th century. By 1850, steel needed for edge tool production was only a tiny component of the steel market. The multiple strategies used to produce steel before the era of bulk process manufacturing technologies are poorly documented and now nearly forgotten.

Steel made from decarburized cast iron (German steel) was by far the most important source of steel in colonial New England in the 17th century. It is now impossible to document the extent that the small quantities of steel derived from blacksmith fined pig iron made in colonial blast furnaces supplanted imported steel. Natural and forged and/or case hardened steel would have provided only a small percentage of steel used during this century. Winthrop's fleet brought more than settlers to the Massachusetts Bay Colony. All these techniques would have been well known to the blacksmiths and shipsmiths who came to New England in the great migration. Knowledge of how to make steel by the continental methods of decarburizing cast iron or by fusion and the awareness of the new cementation process that followed were among the most important contributions to colonial life made by migrating colonists.

In the 18th century, the situation is reversed, with English production of blister steel dominating German steel as the principal source of the steel used by New England's shipsmiths and edge toolmakers. Production of natural steel and forge welded tools would continue to occur, especially in isolated areas, but would account for only a small percentage of the tools made in New England. The mystery remains: to what extent did colonial cementation steel furnaces and steel made by decarburizing cast iron in domestic fineries and forges supplement imported English and German steel after 1713? One other strategy for producing steel remains to be discussed.

Submergence (Fusion)

Perhaps the most ancient, mysterious, and obscure steelmaking technology is the submergence technique, essentially fusion, erroneously called "cofusion" (Barraclough 1984a). This involved the melting of cast iron into a liquid mass, followed by the submergence of wrought iron bar stock into the melted cast iron, which would then be carburized by the excess carbon in the melted cast iron. Raw steel, bar, or sheet stock was then produced from the bloom of raw steel. Numerous variations of the submergence technique process characterize steelmaking strategies from prehistoric China to the early Renaissance. Among these was the technique of bundling wrought iron bars interspersed with fragments of cast iron and forge welding them into steel. In ancient times, or until the end of the bloomery era (c. 1900), whenever bloomsmiths ran their forges at a high temperature (at or above 1130° C), especially with a fuel rich charge, they would discover rivulets of liquid cast iron draining out of the bottom of their furnaces due to the rapid uptake of carbon from their charcoal fuel by the iron being smelted, which would then have a lower melting temperature and become liquid cast iron. It was the accidental or deliberate production of cast iron that would have been the source of the cast iron used in the fusion process before 1350.

In ancient China at some unknown date, as well as in India, Damascus, probably in Africa, and later in Spain and Italy, iron masters knew that, in an enclosed environment such as a crucible, they could melt cast iron in conjunction with small pieces of wrought iron or other carboniferous materials to produce steel. This process of fusing cast and wrought iron is the essence of the ancient Muslim tradition of making Wootz steel. This knowledge was among the age-old secrets of the alchemy of the forge master. Fusion could also occur in open cast iron vats, where the liquid cast iron itself served as the protective enclosure shielding bundled wrought iron bar stock from the deleterious combustion gasses, allowing the production of steel. This latter technique was used in Renaissance Italy, producing Brescian steel, named after one of the principal locations of its production. At this time, production of blast furnace cast iron was ubiquitous, but its use to produce steel appears to have ancient roots in Near Eastern steel-producing communities (Wootz steel). The role of China in the spread of the knowledge of this strategy is unknown.

The propensity of direct process smelting furnaces to produce blooms of iron, raw steel, and cast iron throughout the early Iron Age in European communities is well documented by the archaeometallurgical analysis of the carbon content of the currency bars traded throughout Europe. The high carbon content of cast iron and the impossibility of forging it into tools and weapons would have made the accidental production of cast iron a most unwanted byproduct of early shaft furnaces. There are no written records to document the extent to which European bloomsmiths and steelsmiths would have accidentally or deliberately produced cast iron as a first step in the production of Brescian steel. By 1350, the deliberate production of cast iron became the dominant mode of producing iron in large quantities, facilitated, in part, by the minimal loss of iron ore in the smelting process and high fuel efficiency. Such cast iron had to be refined (decarburized) to produce wrought iron, malleable iron, or German steel. This mass production of blast furnace cast iron was clearly preceded by the limited production of cast iron in Europe for centuries, if not for millennium. The quantity of cast iron deliberately produced before 1350 for the purposes of manufacturing steel by the process of fusion (Brescian steel) will never be known.

Numerous variations of crucible steel production were used to produce the Damascus and Japanese swords. The many regional variations of steel production by fusion are, as yet, undocumented and may also never be known. The uncertainties surrounding the production of Wootz steel (Smith 1960) are just one example of our contemporary inability to unravel and explain steel-producing strategies that were never scientifically documented and are now centuries, if not millennia, old. The revival of the crucible steel process by Benjamin Huntsman is a modern adaptation of this basic principal: blister steel is reheated in conjunction with small quantities of a mixture of carboniferous materials in the enclosed environment of a crucible. In 1855, Sweden lifted its restrictions on exporting its famed low sulfur low phosphorus cast iron to England. The availability of this high quality cast iron coupled with recent advances in understanding the chemistry of ferrous metallurgy led Sheffield and American forge masters to realize that they could omit the step of producing blister steel to make crucible steel. Instead, they utilized what they may have thought was an innovative trick of mixing high quality Swedish or Adirondack charcoal iron with Swedish or American low sulfur cast iron and scrap steel. High quality crucible steel was the result, but did they realize this was a technique that was 2000 years old?

By the time of the building of the *Virginia* at Fort St. George, 1607, knowledge of the Brescian method of making steel would have been as universal among Europeans, including English forge masters and blacksmiths, as was the making of natural steel, the most difficult of the three steelmaking strategies of the time. Nonetheless, making steel by the continental method of decarburizing cast iron (German steel) was the dominant steel-producing strategy of the 16th and 17th centuries and required finesse of judgment about the texture of the bloom of iron evolving from the decarburized melted cast iron. Soon, the coming age of blister steel would replace bloom texture with iron and steel fracture and color as a rule of thumb guide to the production of steel. Meanwhile, any back street Bristol or countryside Forest of Dean/Sussex Weald blacksmith could make raw steel in his forge with small quantities of waste cast iron and wrought iron bar stock. As with other early Iron Age forge masters and blacksmiths, they didn't leave a written record describing their work or their steelmaking strategies. But the legacy of the

68

multiple steelmaking strategies they used lives on in America in the form of the edge tools that survive from the colonial era and the early republic and the shipbuilding industries they facilitated, one of the first examples of which was that icon of icons, Maine's first ship, the pinnace *Virginia*.

In the decades after the construction of the *Virginia*, American colonial era forge masters, shipsmiths, edge toolmakers, and gunsmiths played an important role in the coming revolution. Without their metallurgical skills, craft traditions, and tenacity, and without the rich natural resources at their disposal, the American Revolution would have had a different outcome. A more detailed review of the steel- and toolmaking strategies and techniques in use at the time of the construction of Maine's first ship the pinnace *Virginia* opens the way to an understanding of how the early modern traditions of the indirect process of steelmaking associated with the blast furnace evolved concurrently with the continued use of more traditional direct process bloomery iron production. The simultaneous interaction of both iron-making traditions enabled a robust colonial economy based on shipbuilding, fishing, timber harvesting, bog iron-smelting, local markets, and inter-colonial trade to evolve into the vigorous market economy of the post-Revolutionary era. The prosperous years of the neutral and transoceanic trades and rapid westward expansion had its hidden and nearly invisible enablers: the shipsmiths, toolmakers, and shipbuilders who helped build a nation where the American factory system and the railroads that distributed its products were just over the horizon.

The Hidden Role of Carbon, Silicon, and Manganese

Any review of the many steelmaking strategies used to make the edge tools of the shipwright must acknowledge that edge tools were only one component of Renaissance era (or any era's) toolkits. Gun, sword, and knife production have a much greater historical visibility than the more easily overlooked forging of edge tools or horticultural implements. The history of sword cutlery, for example, is a complicated and, as yet, undocumented and untold labyrinth of forging secrets. Unraveling the relationship between the metallurgy of forged steel edge tools and the metallurgy of malleable iron agricultural tools in European communities helps us understand the subtle differences in the wide variety of tools made by colonial edge toolmakers, blacksmiths, shovel makers, gunsmiths, or sword cutlers. In the early colonial period, both the chemistry and the hidden role of carbon, silicon, and manganese were unknown to the many ironmongers, who migrated to New England, despite their vast empirical knowledge of the diversity of strategies and techniques for making steel and forging tools.

No implements other than weapons (swords and guns) were more important to agrarian Europe than agricultural implements. Ductile wrought iron was too soft for hoes, hay forks, mining picks, shovels, or ox and horse harness hardware, which required iron with a higher carbon content. The "ferfort" produced by the continental method of decarburizing cast iron was harder and more durable and less subject to bending and deformation than pure wrought iron. Some agricultural implements required a steel cutting edge, especially sickles, scythes, and hay knives. Austrian scythe makers were already famous in 1607 for the quality of their steel scythes when the *Virginia* was constructed with steeled edge tools. When European toolmakers made steel using the continental method, which we now call German steel to differentiate it from what later evolved as the English methods of making blister or cast steel from re-melted blister steel, they compensated for their lack of knowledge of role of carbon, silicon, and manganese in steelmaking with their finesse and their ability to determine the quality of the steel they produced by its color, texture, and fracture.

In the case of continental or English forge masters fining cast iron to produce steel, wrought, or malleable iron during the 16th century, correct analysis of the texture of their bloom was essential. High carbon cast iron melts at a relatively low temperature compared to low carbon wrought iron. As the cast iron melts, carbon is removed from the iron by the oxidizing fire, gradually increasing the melting temperature of the iron. When carbon is removed from the now liquid cast iron, a pasty but relatively solid bloom of wrought iron forms, which can be removed by tongs from the finery furnace or later from the puddling furnaces. This bloom of wrought or malleable iron represents the final stage in refining cast iron. Ideally, the resultant product is pure wrought iron with a carbon content of only 0.02 or 0.03%. Due to the tendency of the furnace fire to recarburize the cast iron being fined, the production of pure wrought iron was among the most difficult tasks of the smelter, at least until the perfection of the reverbatory puddling furnace circa 1784. In between the formation of liquid cast iron and the pasty, but solid, bloom of wrought iron is a stage where a slightly less solid, mushier bloom of malleable iron can be manipulated and drawn out of the furnace with a variety of tools and tongs. In almost all situations, the products of the finery shaft furnace were sent to another furnace (called a "chafery" in America) for further mechanical (hammering) and re-heating prior to additional forging to remove siliceous slag contaminants before being forged into iron bar stock. Early bloomsmiths could easily see the silicon fibers in wrought iron, even though they didn't know the name of the element that played such an important role in determining the physical characteristics of tools they were forging. The fining process not only removed slag but also slightly increased the carbon content of the wrought iron, producing charcoal iron with its slightly altered physical qualities of increased hardness and decreased ductility, i.e. malleable iron. This bar stock iron of varying carbon content was fashioned into tools by the blacksmith or sent to other mills for further processing. For partially decarburized cast iron, the raw steel, then called "malleable iron," with a carbon content of 0.2 to 0.5% underwent similar treatment, but the end product was not iron bar stock but a raw low carbon steel bar stock. Most of this "malleable iron" was used to make tools and implements, which today are made from bulk process low carbon steel but without the useful silicon slag inclusions of malleable iron. In some situations, steel with a higher carbon content (> 0.5 cc) than malleable iron was deliberately produced. This steel bar stock had sufficient carbon content to be reforged, quenched, and tempered into steel bars that were then used by the blacksmith and shipsmith to "steel" edge tools. This steel didn't have to be imported from England if knowledgeable forge masters knew how to make it in a colonial furnace by fining cast iron.

In the modern era, we look back on this indirect process-derived bar stock, little of which survives today, and think of it as wrought iron with a low carbon content, essentially the same product as produced by the older tradition of the direct process smelting of iron in a bloomery. Direct process-derived bloomery iron, however, had a higher slag content than the wrought and malleable iron produced from fined or puddled cast iron or bar steel produced from decarburized cast iron. The higher slag content gave direct process bloomery-produced wrought iron a distinctive fibrous appearance; the silicon filaments it contained were helpful for some applications such as anchor forging, and for hardware and tools (e.g. harpoons) requiring a high tensile strength. In general, the lower slag content of fined or puddled wrought or malleable iron, which still contained a significant, visible siliceous slag content, was preferred in most industrial and toolmaking operations. A minimum of slag content, but not its absence, was essential in bar steel used for making steel edge tools of any kind.

The carbon content of both direct process bloomery iron and indirect process puddled iron varies much more widely than our modern categories of "wrought iron," steel, and cast iron suggest. Metallurgical analysis of iron implements made from the currency bars produced by early Iron Age bloomsmiths in continental Europe indicates that such iron had a varying carbon content, according to the intended use of the iron being produced. Even today, there is a continuing confusion about the exact definition of steel. Iron containing less than 0.5% carbon cannot be as easily quenched and hardened into steel as can iron with a carbon content of 0.5% or more. There is, nonetheless, a significant difference in a shovel or hoe made from iron with a 0.4% carbon content, essentially low-carbon steel but still containing significant silicon slag filaments than a less durable one made from pure wrought iron with 0.02% carbon content. Gordon (1996) clearly defines steel as iron having 0.2% or more carbon content. In his glossary, he does not mention the quenching threshold of steel, which must contain 0.5% carbon to be easily quenched and hardened.

In the years before bulk process steel, which didn't appear on the market in large quantities until 1870, significant traces of silicon existed in most ironware (more in bloomery iron, less in puddled iron), providing useful qualities of tensile strength, ductility, and softness that have been lost in the post wrought iron era of bulk processed blast furnace-derived low carbon steel. Ironically, modern "low carbon steel" has the same carbon content, and, in many cases, the same uses as the more convivial "malleable iron" (0.08 – 0.2% carbon content), which had the advantage of containing silicon, which made wrought and malleable iron so much more useful and aesthetically interesting than low carbon steel. Unknown to early toolmakers was the fact that manganese, which provided qualities of hardness and durability, was also an important microconstituent of wrought and malleable iron, though in widely varying quantities. It was also, in the form of spiegeleisen (manganese-laced cast iron), a key alloy in bulk processed steel (after 1870). Already noted as a key ingredient in the ores of Mount Erzberg (Ore Mountain) and the natural steel products of ancient Noricum, manganese is a common microconstituent of bog iron ores, though present in lesser amounts than in the spathic ores of Carinthia and Styria in Austria. The role of manganese in early steel- and toolmaking strategies and techniques links the efforts of early Iron Age and Roman ferrous metallurgists to the later evolution of a robust iron industry in England in the years just before the exploration and settlement of North America.

Before the era of bulk process steel, any iron, even that with a carbon content approaching 0.5%, which could not be hardened by quenching, was called "malleable iron." Our contemporary assertion that malleable iron should be defined as slag bearing iron containing from 0.08 to 0.2% carbon content is one more example of the changing meaning of metallurgical terminology over a period of decades, if not centuries. That early ironmongers also called iron with 0.2 to 0.5% carbon content malleable iron only adds to this disconnect. Metallurgic analysis of many tools and other iron artifacts surviving from the 1650 – 1850 period and now considered to be "wrought iron" would probably reveal carbon content within the range of 0.1 to 0.5% cc, well above the definition of wrought iron (< 0.08 cc), and the ubiquitous presence of both silicon and manganese as microconstituents.

Carbon, silicon, and manganese were not the only elements influencing the metallurgy of iron before its widely varying chemical structure was finally understood in the late 19th century. No ore mined in any location has the same mix of microconstituents. One common contaminant in iron ores is phosphorus. Present in bog iron, it facilitated nail-making but was not useful for most forging purposes. The presence of sulfur in smelted iron ores had and has no useful purpose. Though always present in trace quantities, any significant amount of sulfur (e.g. above 0.05%) results in the phenomena of hot shortness where due to contaminated microstructural formations, hot iron or steel containing sulfur cannot be successfully forged due to a lack of durability, strength, and cohesiveness. In contrast, the presence of manganese in iron not only played a key role in the success of Roman and German Renaissance era metallurgists, it also appears to be an important constituent of ironwares produced in 16th century and early 17th century English iron smelting centers.

As noted, the Forest of Dean and the Sussex of Weald were the two important 16th century centers of iron and steel production in England. Cleere (1985) has observed that, in 1520, at least two blast furnaces were operating in an area that had easy access to the most important toolmaking and shipbuilding center in England, London, which was a short distance to the north. By 1548, a total of 53 blast furnaces, fineries, and bloomeries were operating in the Weald in the waning days of Henry VIII. By 1574, the market economy of the Elizabethan era was beginning to blossom. Cleere notes 52 blast furnaces and 54 fineries and bloomeries in operation. At this time, England faced its fuel crisis, and the deleterious impact of charcoal production resulted in limitations on cutting white oak and other large trees throughout southern England in 1558, 1581, and 1585. Fuel for blast furnaces was often restricted to coppice grown on the great estates of the landed gentry, who had an intimate role in iron and steel production in England at this time. The sources and chemical composition of the ores used by English ironmongers played a key role in the success of this industry.

One of the most interesting aspects of 15th and 16th century iron and steel production in the Weald pertains to the iron ores available at that time. The most important Weald iron ore was a clay ironstone containing ferrous carbonate $FeCO_3$ (siderite). Unlike most other ore beds in England, Sweden, or northern Europe, the ores in Sussex also contained significant quantities of manganese carbonate $MnCO_3$ and magnesium carbonate $MgCO_3$. As with the manganese-rich ores of ancient Noricum, the center of Roman ferrous metallurgy, both the manganese and magnesium carbonates served as a flux during the production of natural steel in the bloomery and the decarburization of cast iron in the finery to produce steel. These fluxes served the useful function of melting at a lower temperature and more efficiently draining off unwanted sulfur, promoting a more uniform uptake of carbon in an iron bloom than would otherwise occur in the attempt to make steel from ores with a different chemistry. Such attempts using non-manganese-containing iron ores were universally less successful and produced steel of inferior quality. Hence the fame of the Noric steel produced in Austria, transported south to Italy through alpine passes, and supplied to Caesar attacking the Gauls (c. 58 – 51 BC) with their greater number of armed warriors and their inferior steel swords. The famed Venetian steel of the 17th century mentioned by Moxon ([1703] 1989) (see Appendix E) was probably derived from these Styrian ores, though steel production in Venice may also have been influenced by Wootz steel technology from Muslim societies to the east, also noted by Moxon as one of the important sources of steel in the 17th century.

The manganese carbonate ores of the Weald played a central role in the evolution of a vigorous English steel industry before the rise of Newcastle and Sheffield as England's major steel-producing centers. Its most important products were weapons and ordnance. Though Henry VIII imported German blacksmiths to make his armor, the steel he used was probably made to the south of London in the Weald of Sussex rather than imported from Germany or France. Henry's attempt to make armor of equal quality to the famed German product failed, possibly because he was too impatient to allow the German smiths sufficient time to anneal the armor they made. Nonetheless, the Wealdean ores available to German smiths working in England at this time had a similar chemistry to the famous "steel ores" from the Erzberg (Ore Mountain) in Austria.

Perhaps due to both the shortage of fuel and suitable iron ores, iron and steel production in the Weald quickly declined after 1600 and, instead, spread to the north, throughout the midlands, and then to the river Derwent and Newcastle in the far northwest. Barraclough (1984a) notes this shift of the British iron and steel industry to the north and the anomalous appearance of Newcastle as England's most important steel-producing center between 1700 and 1775. Though an ancient center of cutlery production and a small steel-producing center during the early 18th century, Sheffield did not surpass Newcastle in steel production until after 1775 (Barraclough, 1984a). At this time, Watts' steam engine greatly increased blast furnace production facilitating the great age of the English steel industry 1775 – 1875.

When the anonymous edge toolmakers forge welded the edge tools used by the shipwrights who built the pinnace *Virginia*, 175 years before the beginning of the hegemony of the Sheffield-based steel industries, they almost certainly used the superior quality steel made from fined cast iron in the Weald. This "German steel" in turn, almost certainly contained significant amounts of manganese as a microconstituent of English ore deposits that are now entirely depleted and forgotten.

German versus Blister Steel

Figure 29 Early English style hewing ax. In the collection of the Davistown Museum 102904T2. There is a probability that this ax was made in Massachusetts in the mid- to late 17th century at Saugus Ironworks, near the location of its recovery (Brookline, MA) rather than in England.

A visual inspection of edge tools still surviving from the 17th century, almost always forged from German steel rather than natural steel, reveal cutting edges, especially on axes, including trade axes, with a different look than the typical 19th century weld steel ax. Edge tools that predate the introduction of the cementation furnace to produce blister steel often lack visible evidence of a steeled cutting edge, as exemplified by the early English style hewing ax shown in Fig. 29. Found in a New England tool collection, this ax may have been made in England but could also have been produced at an early colonial forge site such as the Saugus Ironworks by toolmakers, such as Joseph Jenks, from pattern-molded decarburized cast iron. Such a toolmaking technique would have created a one piece edge tool reminiscent of Bronze Age cast edge tools. This tool can also be considered a precursor of the modern one piece cast steel ax.

Many early modern French and English axes (1450 – 1650) show no sign of a welded steel cutting edge. The extent of this alternative edge toolmaking technique of molding or casting remains undocumented but may have been an important alternative strategy for making edge tools with German steel. Particularly intriguing is a reference in Briggs (1889), where the author notes the presence of an early cast iron hatchet mold from Lambart Despard's 1702 blast furnace on the North River. When that facility was abandoned in the early 18th century, the iron hatchet mold was incorporated in the fire back of the Barker House in Pembroke. The house was left to rot in the early 20th century and an important clue to the toolmaking strategies of early New England was lost. Mechanical (hammering) and thermal (heating) treatment of the cutting edge of such cast tools would have been required to produce a sharp edge tool. The survival of hundreds, if not thousands, of "French trade axes" made in the Biscayne region of Spain, transported over the Pyrenees, and exchanged for furs by French traders before ending up in American tool collections attests to the widespread use of this alternative toolmaking technique. The wide array of edge and other tools made exclusively from German steel (decarburized cast iron) on display at the Maison de l'Outil museum in Troyes, France, is further evidence of the dominance of German steel in continental Europe from 1400 to

1900. A large minority of such trade axes as well as almost all of the axes at the Maison de l'Outil show no visible evidence of the traditional weld steel forging techniques (Fig. 31, sandwich pattern insert and steeled bit, Fig. 32) used to make axes beginning in the early Iron Age. After 1700, steeling edge tools with blister steel dominated English tool production as well as our awareness of how tools were made prior to the era of the all cast steel ax or edge tool.

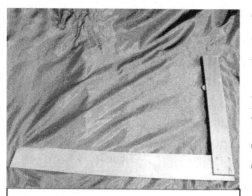

Figure 30 This square is signed "TURNER & __". The second name is not legible. It was made in the 1840s in Massachusetts. It is also marked "German steel" and is probably made from reforged blister steel rather than decarburized cast iron. In the collection of the Davistown Museum 052107T1.

The production of steel by the continental method of fining cast iron was well established in Europe before the first appearance of the blast furnace in England in 1496, or its possible earlier appearance in the Forest of Dean. Shortly after this date, French blacksmiths began immigrating to the Weald; they were almost certainly a principal source of knowledge about the continental method of making steel by decarburizing cast iron, and they may have played a role in the rapid spread of blast furnace technology throughout the Weald and the Midlands after 1520. Barraclough (1984a) describes the later appearance of German smiths working at Shotley Bridge on the river Derwent in 1686. He notes that they were still making steel by the "old German" method. Ironically, in 1686, the first converting furnace for producing blister steel was already in operation in this area, and the rapid conversion to a new more efficient technique for producing steel out of iron ores that did not contain manganese dominated English steel production until the mid-19th century. Use of the converting furnace also quickly spread to Sweden and Danzig, an important Baltic Sea steel-producing center. Benjamin Huntsman demonstrated in 1742 that cast steel of the highest quality could be made by re-melting blister steel in Stourbridge clay crucibles. Sheffield edge toolmakers were soon making those fine quality carving tools and drawknives that we find in American tool chests today. Such tools played a key role in the triumph of the art of making case furniture in the late 18th century in locations such as Newport, Boston, Salem, Portsmouth, and Philadelphia. No such cast steel edge tools were available to the shipwrights at the Popham settlement; nonetheless, the quality of the steel

Figure 31 Axes hanging in a display at the Maison de l'Outil museum in Troyes, France.

in the edge tools used at Fort St. George was still probably superior to that offered in any 21st century American hardware store.

Steel made from fined cast iron was raw steel, as was "natural" heterogeneous steel made from the bloom of the direct process shaft furnace when the fuel to ore ratio was altered, with less ore and more charcoal fuel to carburize the iron. Under either circumstance, production of edge tools, whether socket chisel, scythe, or knife, was contingent upon the skills and talents of the blacksmith who had to reforge the steel for its intended use. In 1865, the Sanderson Brothers, English steel producers and importers, issued the price list (Fig. 27) which notes a wide variety of steel available before the era of bulk steel production. The origin of many of the types of steel in this listing predate the modern era of cast steel and the rise of England's famous iron and steel industry after 1775. "German steel" is a term that appears in this price listing, and German steel in the form of decarburized cast iron dominated the European steel market after 1400. German smiths migrating to England played such an important role in steel production that there still exists some confusion as to the meaning of "German steel." When the German smiths were working at Shotley Bridge on the River Derwent or, 150 years earlier, in the Weald to the south of London, they used steel production techniques that were centuries old. After 1680, the German smiths, however, quickly adapted to the presence of blister steel, which could be produced in greater quantities than by fining cast iron. They were so adept at piling, bundling, and refining blister steel into spring and shear steel that, even today, the high quality steels produced by the reforging of blister steel are still incorrectly called "German steel" and are often so labeled in English, but only on English-made tools of the period. Tools labeled "German steel" in English, including back saws, squares, and other tools, are usually made from shear or double shear (sheaf) steel, i.e. refined blister steel. The Davistown Museum has a number of such refined blister steel tools in its collection, marked "German steel" in English (Fig. 30). Barraclough (1984a) notes the multiple ambiguous meanings of "German" steel. This term should now, in fact, denote steel produced by fining or decarburizing cast iron and not steel made by piling and reforging blister steel. For England and some locations in northern Europe, the converting furnace for the production of blister steel represented a great advance in the quality of the steel being produced. The lack of iron ores containing significant trace amounts of manganese may have played a key role in the rapid adaptation of blister steel production technology in England after 1680. Many efforts at using the continental method of fining cast iron to make steel failed because the iron ores being smelted did not have a chemical composition e.g. a significant manganese, as well as a low sulfur and phosphorus content, compatible with making high quality steel for edge tools and weapons by this process.

The Special Case of Cast Steel

In 1742, Benjamin Huntsman, a Sheffield manufacturer of watch springs, dramatically expanded the options for forging steel for edge tools in his search for better quality steel for his products. As a metallurgist, Huntsman was certainly aware of the mystique of Wootz steel, which had been imported to Paris in the 17th century from the Levant. Attempts by French blacksmiths to reforge it without additional thermal treatment were unsuccessful. Its high carbon content of +/- 1.5% made its use for better quality swords impossible, which was probably the objective of their efforts. De Réaumur (1722) later wrote about cast steel and its potential usefulness and his writings were probably familiar to Huntsman. Barraclough (1984a) notes the earlier production of all-steel swords by the Vikings, who obtained access to Wootz steel in their wide ranging trading routes.

Until the mid-19th century, the role of carbon in facilitating or hindering the forging of steel was unknown. French smiths did not understand why their imported Wootz steel, a form of cast steel whose chemistry was still not clearly understood at the end of the 20th century, could not be repiled and reforged in the time-honored tradition of the blacksmith as sword smith or edge toolmaker. Huntsman was able to replicate a form of Wootz steel by taking high temperature resistant Stourbridge clay crucibles and mixing a few kilograms of broken up blister steel bar stock with carboniferous material such as charcoal. After firing in a furnace for a few hours at a high temperature (1500ºC), the blister steel would melt, and the heterogeneous distribution of carbon characteristic of blister steel would become the homogenous distribution of carbon characteristic of the fine grain high quality cast steel still sought by all timber-framers and other artisans.

Other English steel producers soon became aware of the secrets of Huntsman's re-adaptation of ancient cast steelmaking techniques, and cast steel was quickly adopted for the manufacture of razors and edge tools. Carving tools of the highest quality made by Sheffield and Birmingham manufacturers soon became available to American cabinetmakers, and cast steel as the highest quality weld steel also became available to American edge toolmakers. Cast steel then became an alternative, if more expensive, option to the continued use of sheaf steel and possibly German steel to forge the larger chisels, slices (slicks), and broad ax blades used by New England shipwrights. Still sought today on old edge tools, the imprint "cast steel" advertised them to shipwrights and other woodworkers as the finest available. It was another 80 years before American steelmakers in Pittsburg were able to duplicate the quality of Sheffield cast steel in high-temperature resistant plumbago (lead) crucibles recently invented by Joseph Dixon (1845).

The availability of this cast steel to New England's edge toolmakers and shipwrights, combined with the amazing ability of a few edge toolmakers to tediously reforge piled blister steel (sheaf steel) into edge tools of equal quality to English cast steel tools played a major role in easing the handwork of shipbuilding. The fine quality edge tools made of cast or sheaf steel facilitated the great florescence of American shipbuilding (1790 – 1857). The panic of 1857, the coming of the railroad, the invention of petroleum at Watertown, MA, the appearance of steam-powered machinery in larger shipyards, and the development of the factory system mark 1856 as the beginning of the end of the age of the wooden ship.

Ironically, this decade also records the beginning of the brief reign of America's edge toolmakers utilizing mostly American made cast steel to make the finely crafted cast steel edge tools still sought by woodworkers today. Both Thomas Witherby and the Buck Brothers established their edge tool manufactories in this decade. The Underhill clan was also active in southern New Hampshire at this time and had been making "cast steel" edge tools for decades, probably initially using imported English cast steel. With the era of the wooden ship coming to an end, the continuing availability of high quality cast steel edge tools, in conjunction with the appearance of steam-powered mills and band saws, ushered in a final four decades of the finest sailing vessels ever built, even as steam packets, freighters, and steam-powered navies made their ever increasing appearances on oceanic horizons.

The last consideration in the review of the ferrous metallurgy of New England's shipsmiths and edge toolmakers involves not the various strategies for making steel but the special case of making wrought iron – not just any iron, but nearly carbide-free wrought iron with a carbon content of $<= 0.03\%$. It was this high quality wrought iron, refined to contain a minimum amount of silicon slag that was produced in Sweden and was the first choice of American and English blister steel manufacturers. After 1784, Henry Cort's redesign of both the puddling furnace and the rolling mill allowed the bulk production of relatively large quantities of high quality wrought iron. The history of as well as the problems associated with direct process wrought iron production are an important component of a review of the ferrous metallurgy of New England's shipsmiths and edge toolmakers.

The Myth of Wrought Iron

The forgoing review of the wide variety and ancient origins of steelmaking strategies will hopefully counter the popular view that no steel production occurred during New England's colonial period, let alone throughout the Iron Age prior to the settlement of North America. There is a common misconception that, before the advent of the Bessemer process and modern bulk steel production, all tools, with the possible exception of Damascus steel swords, were made of "iron." In particular, this myth holds that all the iron produced before the appearance of blast furnaces and cast iron was "wrought iron" and that after its appearance iron was either "cast" or "wrought." There is also a popular belief that steel was not produced until the innovative production of cast steel by Benjamin Huntsman in the mid-18th century, followed by the manufacturing of bulk processed steel, including slag-free low carbon or mild steel, which occurred in the third quarter of the 19th century.

The direct process production of iron in simple charcoal-fired slag-tapping and/or non-tapped bowl and shaft furnaces characterized all iron production until the appearance of the blast furnace (c. 1350). The blast furnace was much more efficient in converting iron ore to iron, but its high temperatures created iron with a high carbon content requiring refining into low carbon wrought and malleable iron. After 1350, the vast majority of iron tools produced in Europe and a smaller majority of iron tools produced in America before the introduction of the refractory furnace (+/- 1800) to decarburize cast iron into puddled wrought iron were forged from pig iron that had been decarburized in finery furnaces and, in many cases, further refined in chafery furnaces, as they were called in colonial North America. Gordon (1996) specifically notes the use of a Walloon type furnace at the Saugus Ironworks, which was used for fining the cast iron before it went to the chafery. Among the first to use blast furnaces in Europe were the Swedish, who perfected the art of fining and double-fining charcoal-fired cast iron into the malleable iron bar stock so valued by English and American forge masters and toolmakers.

In continental Europe, most direct process bloom smelting of wrought and malleable iron was discontinued during the early Renaissance because of the efficiency of the blast furnace. The vast majority of iron tools in European and especially continental museums, such as the collection at the Maison de l'Outil museum in Troyes, France, are made from decarburized cast iron, i.e. malleable iron or German steel. Few of these tools are "wrought iron" with a carbon content below 0.08%. While we cannot be certain of the carbon content of many of the axes, bill hooks, and other tools at the Maison de l'Outil, many of the tools on display, with the exception of some of the highest quality axes, are not tool steel (=> 0.5% cc) that has been hardened by quenching and tempering. If we could subject these tools to metallurgical analysis they would probably have a carbon content between 0.2 to 0.5%. One of the mysteries surrounding the production of German steel tools is the extent to which the entire bodies of these (ferfort) steel tools were "somewhat" hardened by quenching and tempering (Barraclough 1984a). Whatever the carbon content of the spectacular array of axes at the Maison de l'Outil, the cutting edges of most edge tools would have been the subject of additional mechanical heat treatment, whether or not their malleable iron or steel bodies had been quenched and tempered. Whatever their carbon content, most working tools, whether in Europe or American, were not made from "wrought iron." These tools, instead, fall into that interesting and controversial gray zone of the smoky environs of ferrous metallurgy, malleable iron with a carbon content in the range above wrought iron (0.08% cc) and below tool steel (0.5% cc).

In the interim between the appearance of the blast furnace and the refractory puddling furnace, the art of producing wrought iron either from decarburized cast iron in finery and chafery furnaces as at Hammersmith or in continental fineries as in France, Sweden, Austria, and Germany was almost as difficult as producing pure wrought iron in direct process bloomeries, which were much more common in colonial North America in the 17th and 18th centuries because of the simplicity of their operation and the wide availability of bog iron ore, which offset their inefficient smelting of iron ores. Inefficient though they were, direct process bloom smelting furnaces continued to produce significant quantities of smelted iron into the late 19th century, especially in rural areas with significant forest resources, such as Catalonia in Spain and North America. The continued production of charcoal as a fuel long after most English and European ironworks had switched to coke-fired blast furnaces and, after 1784, to coke-fired refractory (puddling) furnaces, allowed many American ironworks to avoid contamination of iron bar stock with the sulfur derived from coke fueled furnaces. The high quality wrought iron produced in charcoal fired American bloomeries was often of equal quality to the fined iron bar stock produced in Sweden from charcoal-fired pig iron and used for making blister steel. The ability of American bloomsmiths to produce high grades of charcoal iron played a key role in crucible steel production in America after 1865 and the concomitant florescence of American edge toolmakers. It is now nearly forgotten that the American colonies also produced large quantities of bloomery-smelted charcoal iron, even though larger integrated ironworks, such as the blast furnace complex in Saugus, MA, which produced pig iron that had to be fined (decarburized) into wrought iron, dominate our memory of colonial iron production.

To obtain the tough fibrous malleable wrought iron, which was so sought after by American colonists for making nails, chain, wire, agricultural implements, and hardware, careful control of furnace conditions by the bloomsmith was absolutely essential. It was this bloomsmith who supplied iron bar stock to blacksmiths, shipsmiths, edge toolmakers, and anchorsmiths for their reforging of this iron bar stock into the products of each of these trades. In particular, the bloomsmith had to maintain careful control of both furnace temperatures and the ratios of the combustion gasses CO/CO^2 to produce either low carbon wrought iron or malleable iron. If the furnace temperature was too hot, rapid absorption of the carbon by the iron bloom would result in the liquefaction of the bloom into unwanted cast iron. Wertime (1961) has these comments on the difficulties of smelting that ideal substance, pure wrought iron:

> The small furnaces to be found in ancient China, medieval Europe, or modern Africa possessed the capability to yield high-carbon cast iron. Any effort to avoid this result and to produce low-carbon wrought iron or steel, by means of a process involving only a single step, became a matter of exquisite human and technical restraint. While the internal functioning of the direct-process furnace has not been thoroughly scrutinized by modern metallurgists, enough is known to say that the restraining efforts were directed to avoiding liquefaction of the iron, which occurs largely through the lowering of melting points by the absorption of carbon. Wrought iron melts at about 1,500° C., cast iron at about 1,200°-1,250° C. In order to avoid absorption of carbon the following controls were applied: the judicious placing of tuyères and directing of the draft; careful charging of charcoal and ore to avoid too intimate contact between them (in some cases the charcoal forming a central column between columns of iron); the use of such quantities of charcoal as would help develop an excess of carbon dioxide and smother the fire; slagging of the smelted iron with a poker; the occasional use of silicious fluxes to encourage oxidation of carbon toward the end of a heat; and a relatively short time of contact between iron and fuel. Under ideal circumstances, reduction took place at about 1,200° C., and the iron, instead of liquefying, separated into tiny crystals. (Wertime 1961, 44-45)

Containing only trace amounts of carbon, these tiny crystals of pure ferrite would gradually coalesce within the slag to form conglomerations, or blooms, of wrought iron. In the inefficient bloomery process, large quantities of iron ore as ferrous oxide were lost as they combined with the silicon in the furnace in the form of ferrous silicate. This silicate, the formation of which was encouraged by the alkalis contained with the furnace charcoal ash, served the useful purpose of protecting, by enclosure, the reduced ferrous oxide (now ferrite, the main constituent, along with silicon, of wrought iron) from additional carburization.

What is written on paper as a relatively simple chemical reaction $FeO + C = Fe + CO$ is, in reality, as Wertime (1961) notes, much more complex. Annoying nodules of high carbon steel would often contaminate the crystal blooms of wrought iron. With increasing furnace temperature the risk of liquefying the bloom by increasing its carbon content and lowering its melting temperature was always present. In his attempt to control the disobedient behavior of wrought iron, the difficult work of the bloomsmith is illustrated by the many surviving currency bars from the early Iron Age. Traded throughout central Europe, these currency bars had a widely variable carbon content and were, in some cases, raw steel. The same difficulties characterized colonial production of wrought and malleable iron in direct process bog iron bloomeries, usually listed as forges in town histories and other documents. The iron produced in these colonial bloomeries was also highly variable in its carbon content. All bloomery-produced iron needed to be reforged, rolled, and hammered to expel unwanted slag. That each and every iron application (nails, wire, shovels, anchors, edge tools, etc.) required iron with a differing silicon slag and carbon content attests to the difficulties faced by the bloomsmith and his customers, i.e. blacksmiths, shipsmiths, anchorsmiths, and toolmakers. Each started with bar iron of varying chemical and physical characteristics, including the bar stock produced from decarburized cast iron. Much of this bar stock had already been refined. Such iron bar stock would again be reforged and refined, often in chafery furnaces, to either lower or increase its carbon content depending upon its intended use. Hammering and rolling always served the purpose of expelling slag and impurities and aligning the remaining silicon slag in the wrought iron as its characteristic, useful, horizontal filament inclusions.

In an important early 20th century treatise on wrought iron, by this time made by adding carefully controlled quantities of silicon slag to low carbon Bessemer steel prior to its decarburization, Aston (1939) notes a one inch diameter bar of wrought iron as having 250,000 filaments of silicon slag constituting 1 to 3% of its weight, but a much larger percentage of its volume (Aston 1939, 2-3).

Sponsor and publisher of the now famous Aston text, A. M. Byers Company was the foremost producer of wrought iron in the early 20th century. Their inventive process allowed exacting control of the chemical composition of wrought iron, which could be produced in blooms of up to one ton. Included in the Aston text are chemical analyses of a number of iron specimens produced in the 19th century and derived from iron bridges and other large constructions where tough, rust-resistant wrought iron was a most valued commodity. This earlier stage of high quality wrought iron production was made possible by Henry Cort's puddling furnace, which could decarburize cast iron, producing the large quantities of wrought iron so essential to the expansion and growth of industrial activity in the first half of the 19th century. This puddling-furnace-derived wrought iron was used in America for the first forty years of railroad construction. Steel rails did not replace wrought iron rails until the perfection of the Bessemer process in the early 1870s.

Prior to the invention of the puddling furnace, under any and all conditions the production of pure wrought iron by the bloomsmith was a difficult art. The art of bloomsmithing did not suddenly end in 1784 with the first appearance of the coke-fired puddling furnace, which, took several decades of use before widespread production of high quality puddled wrought iron evolved. Direct process bloomeries continued in use for a century after the invention of the refractory puddling furnace, which was able to prevent contact between the iron being decarburized and the coke fuel so essential for industrial expansion.

Despite the inherent difficulties of the bloomery process, the continued use of small direct process bloomeries to produce wrought and malleable iron, often of a higher quality than puddled wrought iron derived from fined cast iron, was the probable source of a significant portion of the iron bar stock used for making the blister steel for edge tools in the first half of the 19th century. The simultaneous widespread availability of high quality Swedish charcoal iron in the 18th and early 19th centuries fostered the early dismissal of the role of a vigorous colonial and early American direct process iron industry in the period prior to and during the early years of the perfection of the puddling process. American bloomsmiths were working in the context of over two millenniums of bloomsmithing. This early stage of direct process iron production was uniformly characterized by iron bar stock and muck bars with a widely varying heterogeneous carbon content. The bloomsmith refining this iron bloom for its intended special purposes must of necessity had the ability to judge whether his melted iron had a higher than desired carbon content, which sometimes approached that of raw steel, or whether the low carbon high slag content wrought iron needed to be refined and recarburized into malleable iron bar stock.

No bloomsmith or blacksmith working before the American Civil War would have known about the significance of carbon content, but they were able to judge the qualities of the iron they were reforging by its physical characteristics, especially its texture, fracture, and the "feel" of the iron under the hammer. They were also able to direct their mechanical (hammering, rolling, piling) and thermal treatment of this iron (forging, reforging, possibly even annealing) to produce the specific iron types needed by every trade. The wide variety of strategies for direct as well as indirect process manufacture of wrought iron are reflected in the range of names for this iron in the literature describing the history of iron manufacturing as follows: AOK, Berri, C and Crown, CDG, Danks, Dannemora, Double Bullet, Gridiron, Hoop L, Little S, Old Sable (Russian), Orgrund, Spanish, Steinbuck, and W and Crowns (Barraclough, 1984a). In America, there was Juanita, Tennessee, Salisbury, Adirondack, Batsoto Wrought, not to mention Columbia, Mount Hope, Pricipicio, Rocky Mount (Gordon 1996), and the generic terms "charcoaled," "walloon," "double-refined," "triple-refined," "anchor iron," etc. And this is a list of the varieties of wrought and malleable iron excluding the steels and cast irons. The question remains as to which of these was pure wrought iron with a carbon content of 0.03 to 0.08% or less and which were malleable iron with a carbon content approaching or equaling that of low carbon or mild steel (+/-0.1 to < 0.5%). One wonders what contortions our mythic bloomsmith performed, what sufferings they endured to produce our fabled wrought iron? Did any bloomsmith make pure wrought iron, directly reduced from the ore in a bloomery furnace, without enduring the experience of having a trickle of liquid cast iron run out of his smelting furnace or finding those irritating nodules of raw steel in his iron bar stock? And after he fashioned his wrought iron bar stock or smelted higher carbon malleable iron and/or raw steel in his bloomery furnace, what were the toolmaking techniques used to fashion edge tools out of his products or from the blister steel bar stock made out of wrought iron in the cementation furnace? Contemporary working ironmongers may help us answer these questions as old processes and the traditional techniques of working hot iron centuries ago are revived at the dawn of a new age of sustainable creative economies.

Toolmaking Techniques

The primordial form of all edge tools is the knife. Swordmaking is an ancient and specialized variation of the art of knife-making. The forge welding of woodworking edge tools is another variation of the art of the knife maker. The first knives and edge tools were made of iron, not steel. Such tools were no more serviceable, and possibly less durable, than cold-hammered bronze swords and edge tools. Though the earliest edge tools were sometimes made of un-heat treated (not quenched and tempered) low carbon malleable iron, steel edge tools of various types appeared simultaneously throughout eastern, central, and western Europe from the early Iron Age to the late medieval period.

We can sketch three variations of metal toolmaking techniques from the accidental durable metallurgical remnants of prehistory. The rarest and most difficult to forge was the all-steel edge tool best exemplified by the Damascus Wootz steel sword, not to be confused with the pattern-welded damascened swords and guns that survive today. Smith (1960) provides an interesting summary of Damascus sword production techniques; a brief excerpt of his comments is included in Volume 11: *Handbook for Ironmongers: A Glossary of Ferrous Metallurgy Terms* (Brack 2008). It is rather a stretch to connect the Damascus sword to the New England shipsmith, but there may be a link in the form of Viking traders who brought Wootz steel to northern Europe to use in their superior steel swords. Northern Europeans might have obtained knowledge of Wootz steel and how to reforge it when the Vikings ranged throughout Europe, the Black Sea, and westward to the Indian continent. No examples of woodworking tools made from Wootz steel are known to the author. As noted, when French smiths tried to reforge Wootz steel into swords, its high carbon content made it too difficult to use.

The French trading axes exchanged in such great numbers for furs in the 17th and 18th centuries represent the principal alternative method to either natural or blister steel tool production. The majority of these malleable iron or German steel axes were made from partially decarburized cast iron; toolmakers then further carburized their cutting edges. Chemical analysis of a French trade ax (Biscayne ax) would probably show sufficient carbon content to qualify it as a steel ax, at least according to modern definitions of low carbon steel. Trade axes almost certainly had their edges subject to further carburization as illustrated by type 2B knife form (Fig. 32). Before 1700, this form of ax, if subject to proper edge carburization, appears to have been the next best alternative to the dream of forging a Wootz steel ax (or sword). After 1700, the wide availability of blister steel bar stock coincides with the increasing appearance of trade axes made by the traditional methods of the sandwich pattern insert and/or steeled bit ax production (Fig. 32). In the *History of Woodworking Tools*, Goodman (1964) does illustrate a bronze trade ax form prototype found at Gorbunovo, Urals, 1500 BC. Nonetheless, the techniques of the sandwich pattern insert and the steeled bit are far more ancient than the sudden widespread appearance of the malleable iron or German steel trade ax form in the 16th century. One might wonder why many early trade axes (< 1700) don't appear to be made with the sandwich pattern steel core technique. The answer to this question would be the widespread availability of German steel after 1400, which facilitated the production of an early form of the modern cast steel ax. Such an ax, as noted, would have to be subject to further mechanical and thermal treatments by the ax-maker before being ready for sale and use.

The Viking finesse at edge tool production also may have derived from the earlier florescence of the Merovingian sword-makers of the Frankish empires of central Europe. The Merovingian swordmaking abilities in the era before the reign of Charlemagne (768 - 814) were, in turn, derived from the sophisticated abilities of the Celtic ferrous metallurgists who were the inhabitants and sole ironmongers at Halstadt, La Téne, and Noricum. It is interesting to note that the Goths defeat of the Romans in 400 AD may have been due to the availability of sophisticated pattern-welded, double-edged steel and iron swords made by Celtic forge masters at this time, who were the same forge masters who helped Caesar defeat the Gauls in 54 BC. By 400 AD, the quality of the Roman gladius (short sword) may have declined, and Roman access to the high quality ores of the Erzberg (Ore Mountain, Austria in ancient Noricum) may have been restricted by the new political boundaries of the last days of the Roman Empire.

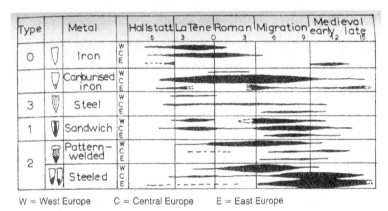

W = West Europe   C = Central Europe   E = East Europe

Figure 7.21   Chronology of knife forms (after Pleiner 1962).

Figure 32 Knife forms. From Ronald F. Tylecote, 1987, *The Early History of Metallurgy in Europe*. New York: Longmans Green, pg. 269. Used with permission from Pearson Education Ltd.

The pattern-welded sword is the most ubiquitous form of edge tool that survived from between the early Iron Age and the late medieval period, the most common edge tool before the era of firearms. Layers of sheet or thin bar steel were interspersed or piled (folded and welded) with thin sheets of wrought iron, reheated and reforged, sometimes many times, to produce strong durable swords. Knowledge of quenching and tempering dates to the early Iron Age and has been verified by metallographic analysis of the crystalline structure of the martinized steel in surviving examples of these swords, which was the product of heat treatment of the swords recovered from archaeological sites in Europe. Wertime (1980) also notes a carefully quenched Egyptian adz dating to the 9th century BC. Existing knowledge of the usefulness of heat treatment dates to the early Iron Age in the Mediterranean region.

Pattern-welding was sometimes also used for knife and cutlery production but not usually for woodworking edge tools. Pleiner's (1962) sketch (Fig. 32) of edge toolmaking techniques from the beginning of the Iron Age to the early modern era remains the best summary of the various techniques used for knife production. Variations in knife making techniques included inserting the steel blade in an iron frame such that continued sharpening of the steel edge would wear away almost the entire knife body. Pleiner calls this method the "sandwich technique." Other techniques included forging of an all steel ax from natural steel, carburizing an iron edge tool in a charcoal fire to steel it, or steeling its edge by welding on a steel cutting edge, also the most common strategy for making woodworking edge tools before the age of cast steel.

In ancient times, that is before the advent of the modern blast furnace (c. 1350) the following four principal techniques were utilized to produce woodworking edge tools:

1. The careful selection or production of nodules or blooms of raw steel in or from the coalescing loupe or bloom in a direct process bloomery furnace, followed by the forge welding of an all steel tool.
2. The slow and tedious carburization of the edge of an otherwise enclosed, e.g. in clay, iron tool.
3. The case hardening of sheet iron in the bed of a charcoal fire to create sheet steel, followed by the folding and welding of this sheet steel into an edge tool. This edge tool was then subject to further forge welding of its cutting edge to homogenize its carbon content.
4. Steeling an iron shaft or poll by welding a piece of steel (the bit) to the iron.

At some unknown time in the past, a knowledgeable blacksmith laminated a thin piece of steel bar stock (the bottom of the chisel) to a softer low carbon steel or malleable iron chisel body, creating a more durable woodworking tool with a harder cutting edge (bottom surface) but less susceptible to chipping due to its softer upper surface. This strategy, a variation of steeling an ax, was commonly used by 19th century edge toolmakers and shipsmiths before the age of the drop-forged all steel socket chisel. A broken socket chisel in the collection of the Davistown Museum (Fig. 33) clearly shows this edge toolmaking technique. The date of the first appearance of this edge toolmaking strategy is unknown.

Only a few edge tools could be produced by reforging nodules of raw steel from the bloomery or by the tedious process of carburizing the lower edge of a chisel. The oxidizing impact of the forge fire probably destroyed many an edge tool during the carburization process. Nineteenth century guides to blacksmithing, such as the one reprinted by Richardson (1978), are full of warnings not to overheat the tool being forged. The systematic production of steel from sheet iron as steel bar stock or at special purpose fineries using raw steel from the bloom probably came to dominate steelmaking strategies in the centuries before the blast furnace simply because larger quantities of relatively high quality steel could be produced by forge masters whose only task was to make steel out of iron. In comparison to modern bulk steelmaking techniques, such sheet or bar steel production strategies were limited to the tedious and erratic daily production

Figure 33 Broken framing chisel seen from bottom, top, and magnified edge. Made of steel, forged malleable iron, and a wood handle with an iron tang, 14 7/8" long with a 4 ½" handle, 1 15/16" wide. In the collection of the Davistown Museum 52907T4.

of a few hundred weight, at most, of steel stock.

As the art of steelmaking became known, sheet and bar steel in small quantities were produced not only for the swordmakers of the Roman Empire but also for the edge tools of the shipwright who built the wooden ships that sailed the Mediterranean and then the Atlantic coast of Europe. The key to producing high quality edge tools with welded steel edges was the extent to which the blacksmith used mechanical (hammering) and heat treatment (quenching then tempering to soften the hard steel) processes on each steel edge tool that they produced. In maritime communities from the early Iron Age to the 18th century, these early edge toolmakers were also the shipsmiths who forged the ironware needed for ship construction (Abell 1948). The many tools they left behind attest to the fact that they immigrated to and worked in the American colonies, not only in New England but also in what was soon to become the most important center of iron production, the Pennsylvania region. The lack of written records in the 17th and early 18th centuries is not sufficient proof of an alleged total dependence on imported English tools by colonial shipwrights. The tools left behind by early American toolmakers provide evidence of a vigorous indigenous edge toolmaking community, which first used imported steel, but, by the mid 18th century, was also using domestically produced steel, often carefully refined from the raw blister steel of the cementation furnace (shear steel).

In the early Iron Age, before the advent of the blast furnace, trade specialization was already evident in the Celtic metallurgical communities of central Europe. The bloomsmith smelted iron currency bars, often with a weight range of 2 to 25 kg and a widely variable carbon content. In some cases, bloomsmiths may have specialized in the production of raw natural steel for sword smiths. The blacksmith had to re-melt the currency bars and forge weld his tools and hardware. Most currency bars subject to metallurgical analysis contain variable amounts of low carbon wrought iron, malleable iron, and raw steel (Tylecote 1987). The smith making edge tools or weapons welded his precious pieces of natural steel onto the iron shaft, socket, or ax poll. In the years after the construction of the *Virginia*, this ancient tradition of toolmaking survived in America not only in the colonial era but throughout rural areas of North America, especially in the Appalachian mountain regions, until the end of the 19th century.

While the nearly atavistic traditions of ancient metallurgy can still be detected in the 18th century in rural areas of Europe and America, a complex market economy of specialized trades was already developing in Europe when the Popham settlement was attempted. Bar steel produced by the sophisticated German finers was probably available as a trade commodity and was certainly the dominant strategy for steel production in England before the age of blister steel. The disastrous Thirty Years War (1618 – 1648) left German industry, including Bavarian and lower Rhine iron-smelting and steelmaking, in ruin, especially after 1634. The interruption in German steel production in this area but not in France may explain, in part, the need for the rapid adoption of the cementation steel process in England and northern Europe. The exclusive use of German steel to produce tools from the 16th to the 19th century in France is demonstrated by the extensive tool collection at the Maison de l'Outil museum in Troyes, France.

By the end of the Elizabethan era, isolated bloomeries making natural raw steel and back alley artisans working up a batch of Brescian steel would have been able to supply only a tiny fraction of the already burgeoning demand for steel for shipsmiths and edge toolmakers who made edge tools for the shipwrights who made settlement of the New World possible. Both the Thirty Years War and the English Civil War that shortly followed greatly increased the demand for steel weapons, swords, and bayonets, which still remained important forms of weapons used for the conventional open field battles of the period. The Great Fire in London (1666) also altered and expanded the need for woodworking edge tool production. Prior to the great London fire, oak dominated timber-framing construction, requiring heavy duty edge tools. England's already depleted oak forests could not meet the sudden post-fire demand for wood. London imported softer lumber, such as fir and spruce, from northern Europe in great quantities. The resulting building boom increased the diversity of molding styles due to the utilization of softwoods, greatly expanding plane forms and designs (Wing 2005). The need for a growing diversity of plane irons following the London fire, using ductile steel more suitable for softwoods, is one of the more obscure metallurgical developments in a century of rapidly changing scientific, political, economic, and geographical landscapes.

George Waymouth's voyage to the Penobscot River area, the building of that first ship on the Kennebec, and the settlement of the north and south Virginia colonies are all landmark events dividing one historic era from another. Soon blacksmiths, often functioning as shipsmiths, would settle in the New England colonies and then in the mid-Atlantic regions to the south. They brought with them a broad knowledge of the ancient traditions of metallurgy at a time of turmoil in England and the demise of the German steel industry, which had flourished in its own Renaissance of watch, lock, and toolmaking for two centuries. German steel production strategies, nonetheless, dominated all continental edge tool production until the mid- or late 19th century. The continued use of the continental method of decarburizing cast iron to produce malleable iron or German steel ("ferfort") is illustrated by the near total absence of either weld steel edge tools or "cast steel" tools at the major repository of French tools at Maison de l'Outil. France became the dominant power in Europe from the early 17th century until the fall of Quebec (1769). The rise of the Sheffield steel industry and the Victorian empire that followed was still a century away. Those first immigrant ironmongers could not have known that the skills, trade secrets, and alchemical finesse that they brought to the American colonies would result in the emergence of an industrial giant centered around the crucible steel furnaces of Pittsburg and the tool factories of New England, New York, and Ohio two centuries later. The building of the pinnace *Virginia*, the first ship built in America, was a landmark event in the adventure of the settlement of North America. The origins and state of contact period ferrous metallurgy, its hidden role in the construction of the *Virginia* and all the ships that were to follow, and in the successful settlement of North America, are nearly forgotten chapters in United States history.

III. The Bog Iron Furnaces and Forges of Southern New England

Shipbuilding in Seventeenth Century New England

Thousands of ships were built along the coast of New England between 1630 and 1700; no record exists of the majority of these smaller vessels. They would have been small fishing craft, coasters, and cordwood boats, ranging from a few tons to 25 or 30 tons, and they were the mainstay of New England's economy. Such small craft would have been built in every cove and small town harbor along the New England coast and would have brought fish, firewood, staves, spars, and other woodenware to larger trading centers such as Boston and Salem. These commodities would have been purchased by merchants who shipped them out to the ports of the southern colonies and the West Indies. The smallest of these coasting and fishing vessels were essential to the survival of every New England community, but they were overshadowed by a vigorous coasting trade whose development coincided with the rapidly expanding New England fishing industry. Goldenberg (1976) makes the importance of Massachusetts and its many shipbuilding communities in the 17th century emphatically clear. While noting the significance of the inter-colonial trade of Massachusetts-built ships that supplied the merchants of Philadelphia and North Carolina, but, oddly, not of Maryland, in the 18th century, Goldenberg comments:

> Throughout the colonial period, Massachusetts merchants and their vessels dominated coastal commerce. Pennsylvania and Rhode Island also had active coasting fleets by the middle of the eighteenth century, but both merchants' correspondence and port records emphasize the unchallenged role of New England as the builder of coasting craft for other colonies. Even Philadelphia merchants occasionally ignored local builders to purchase coasters from New England correspondents. Such purchases must have become common: during the last decade before the Revolution almost 20 percent of Philadelphia's merchant fleet was New England-built. (Goldenberg 1976, 96)

This dominance of colonial shipbuilding begins with and just after the great migration to Massachusetts Bay (1629 – 1643). The first shipping register (Goldenberg 1976) collated by the newly formed colony of Massachusetts describes the types (sloop, ketch, bark, brig, and ship) of ships built in Massachusetts between 1674 and 1714. Sloops, which would include shallops and pinnaces, dominate the register, most averaging 15 to 25 tons. Brigs, the next most common ship listed, averaged 30 to 60 tons. Full rigged merchant ships ranged up to 300 tons but often averaged +/- 150 tons (Goldenberg 1976).

During the 17th century, most of the smaller pinnaces and shallops (sloops) that constituted the backbone of the fishing and timber harvesting trades escaped documentation. Huge numbers of smaller fishing vessels and work boats, such as small cordwood boats for firewood transport and tenders for ship to shore transport, must not have been registered. Baker (1973) records schooners as recently as 1856 being built in Bath as small as 8 tons (the *Hiawatha*), along with sloops and "boats" in the 5- to 10-ton range still being constructed two centuries after most sailing vessels, particularly schooners, were much larger. The shipbuilding records collected by Baker show, however, that registries were still omitting most vessels less than 15 tons. The few small vessels that Baker (1973) notes represent a tiny fraction of those so essential to the efficient functioning of the larger vessels serving the timber harvesting and trading infrastructure of colonial American, the early republic, and the vast trading networks of the 19th century prior to the coming of the railroad.

One might classify the New England shipping community of the late 17th century into four classes according to the size of the vessel as follows:

- Small fishing craft and coasting traders, almost always proto-sloops (pinnaces; shallops; their later cousins, dogbodies; and the Chebacco boat) and usually well under 25 tons, with a working sailing range limited to dozens, and at most, hundreds of nautical miles.
- Coasting vessels, 25 to 50 tons, usually proto-sloops, brigs, and a few barks and ketches, limited primarily to regional trading, with many vessels of the larger tonnage participating in the West Indies trade.
- Merchantmen, 50 to 150 tons, usually brigs and ships, the backbone of the circular trading routes that went to the West Indies, then to Europe and back. Before 1725, most merchant traders were brigs and sloops under 75 tons. After 1725, the schooner design began replacing the brig and brigantines used in the coasting and transatlantic trades, especially for vessels in the lower tonnage range.
- Square-rigged ships, 150 to 400 tons that were used not only in the transatlantic trades but also as transoceanic carriers in the East India trade. These were the ships that went around either Cape Horn or the Cape of Good Hope to go to China and India. Initially owned primarily by Boston and or London merchants, after the Revolutionary War they became the mainstay of Salem's brief heyday as one of the world's most important trading ports.

One of the most interesting chapters in the story of New England's colorful maritime history is the brief ascendancy of Scituate, MA as an important shipbuilding center. For several decades in the late 17th century, it was the third largest shipbuilding center in North America. In the shipping register required by the Act of 1698 for Preventing Frauds and Regulating Abuses of the Plantation Trade and in the Lloyd's Register of

1776, which Goldenberg (1976, 129) used as his primary information source, only Boston (1720 tons, excluding Charlestown 805 tons) exceeded Scituate (1247 tons) in tonnage of ships built. Salem was third at 1173 tons but exceeded Scituate by the beginning of the 18th century. Even in the 1703 - 08 register Scituate is second only to Boston in tonnage of shipping built. Its location on the North River, a rather small river originating in a series of lakes in the nearby towns of Hanover and Pembroke, is a puzzle. Its brief heyday as the second most active New England shipbuilding location speaks to the fact that its upstream forest resources were extremely limited and quickly depleted, yet Scituate continued as an active shipbuilding community for almost two centuries, importing wood and iron fittings from distant locations after being self-sufficient in its early colonial years. But there was also another resource that adds to the mystery of both the river and the vigorous shipbuilding activities that occurred there at the same time that Boston was becoming the most significant port in North America.

The now nearly forgotten resource of bog iron deposits of southeastern Massachusetts played a key role in the rapid evolution of the shipbuilding industry of colonial New England. These bog iron deposits liberated New England shipwrights from total dependence on English deliveries of iron muck bars, often made in Sweden or Russia, and usually carried as ballast in arriving ships. It wasn't all that convenient for a Scituate shipwright to travel to Boston to await the arrival of an English or Boston merchant ship with iron bar stock suitable for forging ships' fittings (and edge tools). Such imported iron bar stock could be transported to smaller shipbuilding communities by coasting vessels, but the availability of locally smelted iron played a major role in the rapid growth of colonial shipbuilding in New England in the second half of the 17th century.

To the south of Scituate, other smaller communities had active shipyards. It is likely that only a minority of ships built at these locations were registered as a requirement of the 1698 Act Regulating the Plantation Trades, which stipulates "fishing craft and vessels trading within the same colony were exempt from registration" (Goldenberg 1973, 129). Rhode Island records for this period are virtually non-existent, but a robust early colonial shipbuilding industry is known to have existed. The extensive activities of ironmongers and forgers in the Taunton River watershed, which drains into Narragansett Bay, also suggest a robust bog iron-shipsmith linkage. Goldenberg goes on to list Duxbury, Plymouth, Manamoit (now Bourne), Rochester, Taunton, Swansea, Bristol, and Rehoboth as early colonial shipbuilding locations (see map Fig. 34). The early Plymouth colony trading post at Aptucxet is the known location of working blacksmith Ezra Perry, whose activities can be dated at least as early as the 1680s (Loring 2001, 3). One of his principal functions would have been shipsmithing, if only for the small vessels built for Buzzards Bay fishing and coasting. All these communities contributed to the vigorous shipbuilding legacy of colonial New England. The actual number of vessels built in these communities, as in Rhode Island, will never be known.

In sketching the story of shipbuilding in colonial New England we again return to the first forges and furnaces at Furnace Brook (Quincy), Hammersmith (Saugus), and Two Mile River (Taunton). It was from these furnaces that the first shipsmiths obtained domestically produced iron for shipsmithing. Ships built in colonial New England prior to the operation of the Saugus blast furnace (1646) were built with imported woodworking tools and fittings derived from imported iron bar stock. The next chapter in colonial shipbuilding extends the vigorous activity of Boston, Salem, and Medford shipyards to a now obscure and forgotten watershed south of Boston that played a surprisingly important role in New England's maritime and industrial history. This particular watershed, that of the North River, is located in the northeastern corner of the bog iron country of southeastern Massachusetts, which was the source of much of the wrought iron used by shipsmiths in the flourishing shipbuilding communities of this area.

The Mystery of the North River Shipsmiths

The environs upstream from Scituate and its neighbors, Marshfield and Norwell, are replete with ponds and bogs, which were very productive sources of bog iron. From the earliest days of colonial settlement, there was a demand for iron for both the shipsmith and the blacksmith who made horticultural tools. The blast furnaces at Quincy and Saugus were only the most obvious and well-documented, though short-lived, consumers of bog iron, which in southeastern Massachusetts at least, was often harvested by flat-bottomed craft equipped with long shovels. The bog ore was roasted in roasting ovens to remove its water content, after which it was an iron ore that was easily reduced in the primitive smelting furnaces of the time. The key questions raised by the vigorous shipbuilding industry of the early colonial period ask not only where they obtained their edge tools but also where they acquired the ironware necessary for the construction of any wooden ship.

The conventional viewpoint is that most or all edge tools would have been imported from England by immigrating shipwrights and woodworkers. The problem of the shipsmith's ironware is more complicated. Given the vast bog iron resources of southeastern Massachusetts and the known presence of bloomery furnaces operating shortly after the construction of the Saugus Ironworks, it is highly unlikely that most of the ironware used for the thousands of vessels of all sizes constructed along the New England coast in the 17th and early 18th centuries, including Scituate, the communities of Buzzards Bay, or the Narragansett Bay archipelago, was imported from England. Each small fishing vessel, not to mention the many larger sailing ships constructed in the late 17th century, required a significant amount of ironware, which had to be forged to fit the specific needs of vessels large and small. Some of this ironware was produced in the many forges that lined the North River to the west of Scituate. By the 1650s, well-documented bloomery iron furnaces had already been established by the Leonard family on the Taunton River and elsewhere and many other late 17th century and early 18th century bloomeries have been described by town historians from Hingham south to Bourne and Dighton (see appendix B). To the south of the lower boundary of the bog iron deposits of Plymouth, Norfork, and Bristol counties, which may be postulated as the Manomet River (Manamoit or Manomoit) of Bourne, now the Cape Cod Canal, isolated bog iron deposits were recorded at Nantucket Island (Sawyer 1998) and Martha's Vineyard (Sawyer 1998; Hine 1908; Banks 1912). To the west of Scituate, the northern range of significant bog iron deposits included the Neponset River drainage, Massapoag Pond in Sharon, and the vast swamps and bogs in Easton, Norton, and Rehoboth that comprised the Taunton River watershed labyrinth. Briggs (1889) notes one forge operating as early as 1650 on the Drinkwater

River, a tributary of the North River, in nearby Hanover. There were likely many more undocumented forges and furnaces scattered along the brooks and rivers of southeastern Massachusetts in locations where there was enough water power to operate a water wheel and trip hammer. Old mill sites are ubiquitous throughout the watersheds of this and other New England regions. We don't know how many of these mills powered small furnaces and forges in the late 17th and early 18th centuries.

One of the most intriguing comments that Briggs (1889) makes is his observation of a blast furnace erected "as long ago as 1702, and leased or hired to a Mr. Lambart Despard for the purpose of casting all kinds of ironware. Hatchets were made here. One casting still in existence is the back of the fireplace now in the old Barker house in Pembroke" (Briggs 1889, 2). It should be remembered that at this early date there is no record of any steel (cementation) furnaces operating in the American colonies. It's highly likely that in 1702, ironmonger Despard was, in fact, utilizing the continental method of decarburizing cast iron to make some of his ironware, especially the hatchet noted as molded in an iron cast. It certainly would not have been a high carbon cast iron hatchet and could only have been cast or molded from decarburized cast iron. Could this fact be confirmed, it would be a most important discovery with respect to how the first colonists in New England might have made some of their edge tools. The mysteries of the North River indeed.

Briggs' (1889) convoluted descriptions of the forges, furnaces, shipyards, and shipbuilders located on the North River and its tributaries indicate that most forges and furnaces had multiple partners, including many members of the same family. As a parent died off, another partner would join the company, as with Barstow's forge, which was operated by Robert Salmon in its later days. But Briggs' survey of activities on the North River retells just a fragment of a much larger story of shipbuilding and shipsmithing on the many rivers in New England and in the colonies to the south. Briggs does not differentiate the smelting furnaces of the bloomsmith, usually located at a water privilege, which could power his trip hammer, from the ubiquitous forges of the village blacksmith or shipsmith, who could work in any location, powering his forge with bellows and forge welding by hand the tools and hardware produced from the bloomsmiths' bar stock.

The following chapters are an ongoing sketch of the iron industries of just one corner of New England, the bog iron country from Hingham south to Carver, the Taunton River watershed (see map Fig. 34), and the Weweantic and Sippican Rivers at the head of Buzzards Bay. A survey of these forgotten forges and furnaces helps fill in the gap between our initial reliance on English tools and iron products and the late colonial era development of a vigorous domestic iron industry in the Pine Barrens of NJ, the headwaters of Chesapeake Bay (Maryland), Pennsylvania, and elsewhere. Many other New England and maritime peninsula locations, such as the Mystic River seaport in

Connecticut, the Merrimac River watershed, the Piscataquis River basin and the St. John/New Brunswick watershed, are worthy of their own voluminous surveys of forgotten furnaces, forges, shipsmiths, and edge toolmakers. A more detailed commentary on the colonial era iron-making industries in southern New England helps link the forgotten shipsmith and his ferrous metallurgic traditions with the vigorous shipbuilding industry of 19th century New England. That first century of activity by southeastern Massachusetts bog iron bloomeries and small blast furnaces played a key role in the success of New England's maritime economy. After 1750, its continuing productivity became a mere detail in the larger narrative of what was already an American iron industry.

But what is bog iron and why was southeastern Massachusetts the location of colonial America's most important iron ore resource prior to the exploitation of the rock ores of Connecticut, New York, Maryland, and especially Pennsylvania after 1720? Along with the story of the shipsmith, the lost history of southeastern New England's bog iron country is another forgotten chapter in America's maritime and industrial history.

The Bog Iron Country of Southeastern Massachusetts

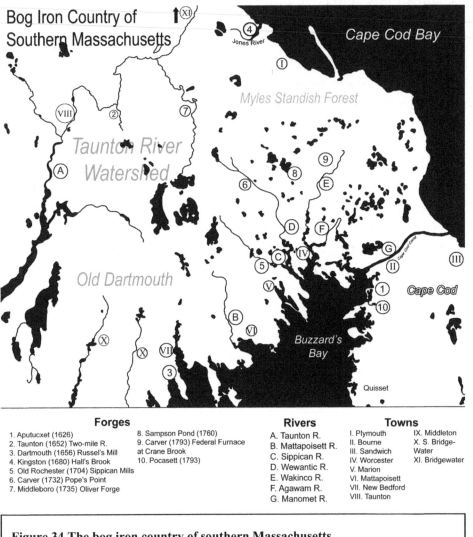

Bog iron is formed when iron oxides in standing and running water are precipitated by bacterial action after contact with oxygen. In part due to the presence of glacial runoff and glacial soils, including granite gneiss, the southeastern coastal plain of Massachusetts was the location of the largest hydrated bog iron deposits known to exist in the colonies until the discovery of the deposits of the New Jersey Pine Barrens. Discovery and exploration of the rock ore deposits of western Connecticut,

Forges

1. Aputucxet (1626)
2. Taunton (1652) Two-mile R.
3. Dartmouth (1656) Russel's Mill
4. Kingston (1680) Hall's Brook
5. Old Rochester (1704) Sippican Mills
6. Carver (1732) Pope's Point
7. Middleboro (1735) Oliver Forge
8. Sampson Pond (1760)
9. Carver (1793) Federal Furnace at Crane Brook
10. Pocasett (1793)

Rivers

A. Taunton R.
B. Mattapoisett R.
C. Sippican R.
D. Wewantic R.
E. Wakinco R.
F. Agawam R.
G. Manomet R.

Towns

I. Plymouth
II. Bourne
III. Sandwich
IV. Worcester
V. Marion
VI. Mattapoisett
VII. New Bedford
VIII. Taunton
IX. Middleton
X. S. Bridge-Water
XI. Bridgewater

Figure 34 The bog iron country of southern Massachusetts.

Pennsylvania, and elsewhere occurred almost a century after the beginning of southern New England-iron-making industries. While the bog iron deposits north of Boston that supplied the Saugus Ironworks were quickly depleted, much more extensive deposits of iron were located in Norfolk, Bristol, and Plymouth counties to the south of Boston and west of Cape Cod. The North River at Scituate and the Neponset River at North Quincy drained the eastern and northern most zones of these deposits. The principal watershed within this bog iron district is the Taunton River, which includes many important tributaries such as the Titicut, Matfield, and Town rivers. To the east, the Jones River and its tributary, the Winnetuxet, also drained bog iron country. To the south, the Weweantic and Sippican rivers enter Buzzards Bay just north of Marion, and the Agawam and

Wancinko empty into the Wareham Narrows. The Manomet River in Old Bourne, once the location of the ancient trading route portage between Massachusetts Bay and Buzzards Bay, is the southeastern most watershed of this distinctive geological glacier-derived resource. Significant deposits of bog iron exist on Cape Cod, especially to the east of Pocasset and on Martha's Vineyard but are not associated with any significant watersheds capable of powering a trip hammer other than the Harlow River at the Pocasset blast furnaces. Beginning in the mid-17th century, dozens of bloomery furnaces and forges were constructed within this unique watershed.

Appendix B contains an ongoing list of the working forges and furnaces that were constructed in this region beginning in the mid-17th century. Many other ironworking facilities will remain forever undocumented, especially those small forges that characterized the typical village blacksmith or isolated farm workshop. Often the only vestiges of these primitive ironworks are the ubiquitous anvils and leg vises that survive to tell us that many a 17th or 18th century farm family made their own tools and that many shipwrights who built their own 3 to 5 ton shallops and work boats may have "ironed" their boats with fittings made from local bog iron by their community shipsmith (i.e. the village blacksmith). In shipbuilding communities, the shipsmith forged not only tools but also the iron fittings for thousands of small vessels that sailed out of New England coves and harbors by the early 18th century. The later shipwrights who built the clipper ships of East Boston or Damariscotta, the downeasters of Penobscot Bay, or the schooners of Bath and Waldoboro labored in the context of two centuries of New England shipbuilding, bog iron harvesting, shipsmithing, and edge toolmaking.

In the beginning of the 18th century, other ironworking centers were established to the south of New England. One important location, which played such an important role in the Revolutionary War, was the New Jersey Pine Barrens and the iron industries (e.g. Batsto Forge) that the bog iron deposits in these barrens engendered. Soon outstripping the production of bog iron both in New England and in New Jersey were the ore deposits in Pennsylvania, Maryland, New York, and elsewhere, which made America the producer of 1/7th of the western world's iron ore by the time of the Revolutionary War. The focus of our listing of forges and furnaces in Appendix B is the vigorous but poorly documented iron industries of southeastern Massachusetts, which played such a critical role in colonial history between 1650 and 1750. It also includes important colonial ironworks north of Boston. Many of the town histories from which this information is extracted contain only the briefest reference to these iron facilities. Of particular importance is the reality that most town historians paid little attention to the blacksmiths, shipsmiths, edge toolmakers, forges, and furnaces in their communities, frequently failing to note their location, dates of operation, or the products they cast or forged. Often only the most long-lived and famous forges and furnaces are remembered in these histories.

Notable examples of well known ironworks include Hammersmith at Saugus, the Leonard forge at Two Mile River in Taunton, Peter Oliver at Middleboro, and Oliver Ames at North Easton. Murdock (1937) provides a comprehensive description of the blast furnaces at Carver without noting the numerous bog iron bloomeries, which would have also been an integral part of the landscape at that time. Also not noted are the small finery forges, which would have reprocessed at least some of the bloomery-smelted muck bar and blast furnace-derived cast iron into tools for local use and hardware for the many shipwrights who worked along the coast to the south in Wareham, Old Rochester, and Dartmouth (part of which is now New Bedford) or along the south coastline of Cape Cod from Woods Hole east to Chatham and north to Eastham. The growth of New England's shipbuilding industry, first in Massachusetts and later in Maine, is intimately tied to the evolution of the ironworking industries of southeastern Massachusetts and the growing colonial ability to forge ship fittings, then edge tools; cast hollowware, then cannons and to make wrought iron flintlock guns for the Revolutionary War.

While the bloomsmiths and blacksmiths of southeastern Massachusetts were busy laying the metallurgical foundations for a new nation, the vast forested wilderness of the maritime peninsula to the east was still unexplored. The forests along the northern New England shoreline quickly became an important timber resource, not only for the masts of the King's ships and the transoceanic India merchantmen but also for the thousands of smaller coasting and fishing vessels constructed in southern New England, which was quickly depleted of timber suitable for shipbuilding, especially white oak. But rather than moving their shipyards to another location, with a few exceptions, most shipwrights continued to work where they lived, importing timber from downeast Maine and from southern colonies. The vast bog iron resources of southeastern Massachusetts played a critical role in perpetuating the longevity of the shipbuilding communities of Massachusetts Bay (Newburyport, Essex, Salem, Medford, Boston, and Scituate), Cape Cod Bay (Kingston, Duxbury, and Plymouth), and Buzzards Bay (Old Rochester and New Bedford). Later, in the 19th century, a vast shipbuilding industry would evolve in coastal Maine; the mystery of the sources of the iron used for their ships' fittings and the steel used to make their edge tools is explored in subsequent chapters. Our sketch of the bog iron industries of southeastern Massachusetts would be incomplete without a consideration of the important role the Leonard clan played in the early growth of America's indigenous iron industry or of the later role of the Carver blast furnaces as a obscure component of this milieu.

The Leonard Clan of the Taunton River

South of the North River, a vast network of bloomeries, forges, and blast furnaces had arisen in the Taunton River watershed beginning in 1652. Some farm families smelted bog iron in forgotten backyard bloomeries and made their own iron tools; others sold the bog iron they harvested to larger furnaces and forges. In turn, these bloomeries and forges supplied the blacksmiths, shipsmiths, and anchorsmiths of the southeastern Massachusetts coastal plain. Iron muck bars from these forges became a form of currency, often paid in hundredweight bars, which could then be traded for food and other necessities (Smith 1983). In 1652, James Leonard came from Braintree to the Two Mile River in Taunton, now Raynham, to begin the hegemony of the Leonard family of forge masters. The Leonard family smelted iron in bloomeries and blast furnaces for almost two centuries at locations in the Taunton, Norton (Easton), and Middleboro region, on the Sippican River in Old Rochester (Marion), and to the north of Boston at Rowley. As late as 1805, four Leonard brothers were reported to be operating an iron mill at Horseshoe Pond in Wareham. They had previously also operated a forge and furnace at Little England in West Wareham from 1798 until 1822 (Chaffin 1886).

Leonard-smelted wrought and malleable iron ironed ships from Wareham and Old Rochester to Quissett and Woods Hole, from New Bedford to Fall River, from Newport, RI to Mystic, CT, and beyond. The first stop for the Leonard iron bar stock (muck bars) was the forges and furnaces of local blacksmiths, many associated with the integrated ironworks of the Leonards, who often refined the slag-laced muck bars of a bloomery (direct process) or blast furnace-derived pig iron (indirect process) to make horticultural tools for the growing settlements of southeastern Massachusetts. A second, often unrecognized, market for bog iron bar stock was comprised of the shipsmiths, most as yet unidentified, who ironed the vast fleet of colonial fishing vessels being built North of Cape Cod, along the shores of Cape Cod and Buzzards Bay, or within the archipelago of Narragansett Bay. Bining (1993) articulates what has been the prevailing view of colonial ironworks after the failure of the Saugus Ironworks:

> On the whole the attempts to establish an iron industry in New England during the seventeenth century were not very successful. Several of the enterprises failed and most of the works were in operation only intermittently. It was estimated that there were only five ironworks in the New England colonies in 1673. A few years later Edward Randolph reported to the home government the existence of six. Many factors contributed to prevent the development of the industry. Numerous lawsuits, Indian attacks, prejudice and opposition because of the large amount of wood consumed by the works, and a preference for English iron, all must be considered among the many reasons which prevented the development of colonial iron manufacture at this time. Not until after the close of Queen Anne's War and during

the years which followed was the iron industry definitely established in the American colonies. (Bining 1993, 13-14)

Almost three quarters of a century after Bining's survey of New England's ironworks, many additional bloomery furnace locations have been documented. Implicit in the booming mercantile communities of Boston and Salem in the late 17th and early 18th centuries and the concurrent shipbuilding boom, which furnished the ships for their far reaching trading activities was the rise of a vigorous domestic ironworking community of tool and ships hardware forging. England and Sweden were not the only source of iron bar stock used by what was already, in 1700, a vast network of community blacksmiths and shipsmiths serving the need of a booming maritime economy.

When Thomas Coram began building ships on the Taunton River at North Dighton in 1697, the Leonard forge located upstream at Two Mile River had already been in operation over forty years. The smelted iron bar stock produced by the Leonard forges and furnaces was so valuable that it became a form of currency and was used to pay the salaries of the schoolteachers of Raynham in the late 17th century (Goldenberg 1976). The iron bar stock fined from the smelted blooms of bog iron was joined by the sows, pigs, and cast hollowware of the blast furnaces that also soon dotted the landscape. A 10,000 year old accumulation of bog iron (limonite) needed only harvesting and roasting to eliminate its 18% water content and ready it for reduction in the many often undocumented bloomeries which spread over the Massachusetts coastal plain. The forges and furnaces of the Leonard family were only the largest and most well documented of a network of water-powered bog iron bloomeries that served the ubiquitous blacksmith and shipsmith shops that characterized every village and shipbuilding community in colonial New England. Anchor forges were also common in this district. Only the gradual depletion of the bog iron deposits and the rapid growth of the iron industries in Pennsylvania, Maryland, and elsewhere after 1720 ended the hegemony of Massachusetts smelters and forge masters in the Colonial Period.

Before 1675, the Taunton River watershed was the most important riverine trade and travel for First Nation Wampanoag peoples and soon became a very convenient avenue of commerce for many inland southeastern Massachusetts iron and ordnance producing communities, which soon nurtured the incipient shipbuilding industry of the Buzzards Bay and Narragansett Bay regions, especially after the end of King Philip's War. The Taunton River was the sole avenue of access to the thriving commercial center of Taunton in the days before the coming of the railroad and the heyday of whale ship construction at Dartmouth including New Bedford, Old Rochester (Mattapoisett and Marion), and Wareham.

When James Leonard left the Saugus Ironworks to establish a forge at the Two Mile River, he brought more than the knowledge of modern smelting techniques he had learned in England. He brought a whole family of ironworkers to establish a series of forges and furnaces that operated as recently as 1876 (Mulholland 1981). His brother Henry had managed the short-lived Quincy ironworks, and his son Thomas was soon a manager at the Taunton (Raynham) works (1683). Henry later moved to New Jersey (1674) to establish the Shrewsbury Furnace. Before going there, he also established iron works at Rowley Village (Boxford, MA) with his son Nathaniel. Edmund Bridges had already established a forge nearby as early as 1636, which would have been a likely source for iron fittings for ships soon to be built at Newbury (Mulholland 1981, 29). It is not known which Leonard established the forge at Sippican Mills on the Sippican River in Old Rochester (+/- 1690) (Ryder 1975).

In his survey of metals in colonial America, Mulholland (1981) is rather emphatic about the failure of the Saugus Ironworks at Hammersmith and later efforts at smelting iron in colonial New England: too large a facility, too complicated, too many owners and managers, and too little funding. He fails to mention that a principal reason for its closure was the depletion of the local bog iron deposits in the area adjacent to the Saugus Ironworks. Mulholland notes that a key reason for the successful longevity of the later Leonard forges was the simplicity of their small bloomeries, which were hardly more complicated than farm forges, often using hand-pumped bellows instead of water wheels, and sometimes serving the dual function of smelting a bloom of bog iron and forging it into primitive tools.

Not mentioned in Mulholland's survey is the vast size of the bog iron deposits, which underlay the success of the Leonard clan. The Leonard clan reflected the influence of a multitude of factors that had a hidden impact on New England's shipbuilding industry in the decades after 1650. In particular, the shortage of iron in England was a principal reason for the establishment of the Saugus Ironworks, which was intended to be a source of iron for both England and the colonists, as well as a source of specie for colonial organizers. In this context, Henry and James Leonard and Joseph Jenks (later Jencks) were only part of a whole contingent of experienced ironworkers brought to the colonies by John Winthrop. James' sons included not only Thomas, but also Josiah, Ben and Uriah; Henry's son was Nathaniel. Later, another "James" Leonard appeared at North Easton along with son Eliphalet. Elkanah Leonard worked one of the Taunton furnaces after 1724, followed by Captain Zephaniah Leonard. Eliphalet's son Jacob was at North Easton circa 1742 and later appeared at Easton, the northern limit of the Taunton River watershed, followed by another son or grandson, Jonathan. Jacob's son Isaac was another Leonard ironmonger who was also at North Easton circa 1792. Jonathan Leonard reappeared at Easton circa 1792 (or perhaps Jonathan Jr.) along with Eliphalet Leonard II and III. Many other Leonards were bog iron country bloomsmiths; their relationships are not clear. In retrospect, the Leonard clan had a widespread influence on spreading knowledge of smelting and forging techniques throughout eastern and south coastal New England. Intimately connected to their circle are famous names such as John King of Kings Furnace (Taunton), the gunsmith Hugh Orr (Bridgewater), Thomas Ames (West Bridgewater, father of James Ames and grandfather of famed shovel maker Oliver Ames), the Captain Joseph Barstow clan, and Peter Oliver of Middleboro, the only known Tory among these ironmongers. Previous to this investigation, historians have forgotten to inquire what role they all played in "ironing" the vast fleet of ships built in colonial Massachusetts, many of which were later sold to English and other merchants as one of colonial New England's most important trading commodities.

The Taunton River was not the only significant watercourse south of the North River, but it was the largest. The ironmongers on the Jones River in Kingston, a much smaller stream to the east of the Taunton River watershed that emptied into Cape Cod Bay, had a later start (1734) and a much shorter run to the ocean. Like the shipwrights of Scituate and Essex to the north, they built hundreds of ships on one small river, few of which appear to be documented in Goldenberg's (1976) survey of colonial shipbuilding. Among the forge and furnace owners and toolmakers on its watercourse was one of the foremost and renowned of all American toolmakers, Christopher Prince Drew. Among the many industries along the Jones River, he was a latecomer (1837) and probably unrecognized in his time as an important toolmaker for the shipsmiths and shipwrights of the early and mid-19th century. The story of Drew and the other Kingston tool (auger) makers is the story of the journey of American toolmakers from the bog iron bloomeries, blast furnaces, and trip hammers of the Leonards et al. of colonial New England to participation in a growing industrial economy based on the gradual appearance of an American factory system utilizing interchangeable parts and a series of innovative machine designs, which appeared in the 19th century. The obscure improvements in the toolmaking machinery at the Drew factory, which his granddaughter Emily wrote about in 1937, provide insight into the growth and success of American toolmaking as the result of the creative efforts of the many entrepreneurs who made New England the center of American hand tool production in the late 18th and early 19th centuries. This legacy began with the first furnaces established by John Winthrop at Quincy and Saugus. James Leonard worked at the Braintree blast furnace during it brief period of operation in 1646, and then in Saugus after 1646, before moving to Taunton in 1652. The forges and furnaces he and his brother Henry and their sons established operated for almost 200 years, much longer than the Saugus Iron Works. His establishment at Two Mile River marked the beginning of a bog iron-based industrial economy that continued with the spread of furnaces, forges, and blast furnaces throughout southeastern New England before developing in the much larger iron producing regions of western Connecticut, bog iron swamps of New Jersey, and the rock ore deposits of Pennsylvania after 1720. Christopher Prince Drew represents the culmination of two centuries of iron-smelting and toolmaking in southeastern Massachusetts that began with the great migration and arrival of James and Henry Leonard. Long before Christopher Prince Drew began his well documented reign as America's foremost manufacturer of caulking irons, mallets, and cats paws, a whole network of furnaces and forges spread throughout southeastern Massachusetts. The Carver blast furnaces are just one link in this now obscure labyrinth, which includes the mysterious Leonard forge at Sippican Mills on the tiny Sippican River. The Weweantic River and its connection to the Wareham nail factories is one more component of the southeastern Massachusetts grid of colonial era and early republic ironworks. In the case of both the Weweantic and Sippican rivers, not to mention Manwarren Beal's Head Harbor Island and Bunker Hole, old but favorite haunts of the author help determine which labyrinths in the historical microstructure of ferrous metallurgy are chosen for exploration in this whimsical survey of the iron industries of New England before 1820.

The Carver Blast Furnaces

The Weweantic River is a tough sail for a coaster. It is a twisting channel littered with glacial rocks and miles of mud flats before the head of tide. In the early days (1685 – 1820), they floated lumber, then bog iron, down its lazy meanders. For the few coasting schooners built upstream in its tidewaters, the Weweantic was a one way trip downstream to the West Indies trade. If any whaling ships were built in the heyday of whaling ship construction (1810 – 1850) in that obscure river, the Sippican, which empties into the Weweantic River, no record of this activity survives. A Leonard forge +/- 1690, was located a few miles upstream from the junction of these two rivers, supplying the shipsmiths who were already at work in Mattapoisett and Marion (Old Rochester) by the late 17th century.

The nearby Wareham Narrows are much more accessible than the Weweantic River. Arriving coasters could sail into the inner harbor and then tack left to the Wareham Narrows, which was a most friendly place to anchor. There are no falls above the Wareham Narrows, just the tidal flats of the Agawam and Wancinko rivers. Coasters couldn't go above these narrows, but the village port of Wareham lined the south bank of the narrows, and the Cape Cod shipbuilding company still builds pleasure craft on the north shore of the narrows, where many a schooner and some New Bedford whalers went down the ways. The road to the blast furnaces of Carver was a straight run to the north. Bar iron from the Carver blast furnaces at Popes Point and Fresh Meadow may have been floated down the Weweantic River to supply the shipsmiths building the first whalers and coasting schooners at Old Rochester and Dartmouth in the late 18th century for the mariners of wood starved Nantucket and Martha's Vineyard. Most of the products of the Carver blast furnaces would have been brought down the Wareham road by oxen to the Wareham Narrows. These products included bar iron for the shipsmiths of Old Rochester and elsewhere and possibly for an anchorsmith or two who worked at Bourne. The most important products of the Carver blast furnaces, however, were cannonballs and hollowware from the Federal Furnace at Crane Brook and the smaller furnaces at South Meadows and the Wenham Brook. Among the largest of the Carver blast furnaces was the Barrows Furnace at Sampsons Pond. The shot and cannonballs from these furnaces were used both in the Revolutionary War and the War of 1812 (Murdock 1937). If hollowware and cannon were floated down the Wancinko River, there is no record of it.

Another Carver blast furnace was the Charlotte Furnace in South Carver, begun by Bartlett Murdock in 1760. Griffith (1913) notes that, as early as 1790, the Charlotte Furnace and other Carver furnaces were facing the gradual depletion of bog iron ores to the extent that "As native ore could not be procured in abundance to meet the demands of the increasing business, Jersey ore was imported through Wareham wharfs" (Griffith

1913, 200). Sometime after 1790, Benjamin Ellis became associated with Charlotte Furnace. Griffith notes that "in 1800 he began to buy shares in the business and by 1808 he owned a controlling interest" (Griffith 1913, 203). The War of 1812 provided an opportunity for Ellis to extend the scope of his business activities in the South Carver/Wareham area. Among the many projects with which he was involved was the re-firing of the blast furnace at Cranebrook, in conjunction with his partnership with Col. Bartlett Murdock, which was now several decades old. Griffith then provides a brief overview of the extent of distribution of Carver-produced cast iron.

> After the war was ended, and with a surplus of capital, the firm was in a position to extend its trade. It began to own vessels through which ore was landed at Wareham, and an extensive teaming business flourished between the plant and the wharf. Vessel loads of ware were also sent up and down the coast from Bangor to New Orleans. The Maine trade thus established continued through the various managements of the plant to the end of the career of Ellis Foundry Company. (Griffith 1913, 205)

Griffith's comments illustrate the continued significance and far reaching distribution patterns of a coasting trade that was well established by the mid-17th century.

Providing an interesting footnote to the coast-wide trade of the early Republic are a chimney cleanout door and cap, cast iron artifacts brought to Liberty Tool Co. in April 2007 after being salvaged from an early 18th century, soon-to-be-torn down farmhouse in the Kennebec River area near Augusta (Fig. 35).

The survival of these two "accidental durable remnants" and the many hand tools and other forged iron artifacts found in New England tool chests and workshops dating from the early 19th century or earlier attest to the extent and longevity of a coasting trade that brought tools and ironware from southern New England and other locations to riverine and head-of-tide destinations throughout the Atlantic seaboard in the era before rail

Figure 35 Chimney door and chimney cap from Ellis, South Carver, Massachusetts.

transport. For the tenacious New England coast-wide trade, transport of goods of any kind around the treacherous Nantucket shoals in any direction was as much a routine component of a sailing voyage as were New England's unpredictable weather patterns. The presence of Ellis furnace-derived artifacts in a central Maine farmhouse built almost two centuries ago illustrates the extent of this coasting trade, which had been established for a century and a half before Ellis shipped his furnace doors to central Maine.

The Carver blast furnaces were only one component of the vast network of forges and furnaces of the bog iron country of southeastern Massachusetts, which achieved its maximum period of production between 1750 and the War of 1812. Just north of the Carver complex were the forges and furnaces of Plympton, and it is probable that many of the products of these nearby ironworks were also brought south to the Wareham Narrows by oxen, following the same route used by the Carver manufactories. In Plympton, two forges managed by Joseph Scott are noted in operation in 1720. Gideon Brand is recorded as having established a cannon factory in 1750 at Dennetts Pond on the Jones River, which empties into Cape Cod Bay at Kingston, a second possible shipping route for items made at this forge and foundry. At an unknown date, but probably within the same time period (+/- 1750), Nathaniel Thomas established an iron forge on the Winnetuxet River. Active as a blacksmith's forge and probable toolmaking location, this forge was still in operation in 1869, where Jonathan Parks made shovels at this late date. There was also a rolling mill just upstream from this forge, where Oliver Perkins was recorded as making nails and bolts circa 1822 (Plympton Historical Society map). These facilities illustrate the longevity of an ironworking industry that was established in the mid-17th century and was still providing ironwares to New England residents and coasting traders, while the steel-producing city of Pittsburg was well under construction to the south in Pennsylvania.

Southwest of the Carver blast furnaces, Old Rochester, now the towns of Marion and Mattapoisett, became the shipbuilding center for New Bedford whalers, just as Essex was the center of shipbuilding for the Gloucester schooners, which were constructed in such great numbers from the late 18th to late 19th centuries. The identity and location of forges, blacksmiths, and shipsmiths who may have fined Carver pig iron into wrought or malleable iron has not been determined, but the Wenham Brook blast furnace later became a cupola furnace and foundry. Numerous cupola furnaces and rolling mills manufacturing nails were working not only on the Weweantic River at West Wareham but just above the Wareham Narrows by the second decade of the 19th century. The most famous of these was and is the Tremont Nail Company, established in 1819, whose facilities can still be visited today. That these industries, most now forgotten, had their origins 150 years earlier in the florescence of a robust early colonial ironworking industry attests to its lingering impact on the industrial growth of the early Republic, especially in areas such as southeastern Massachusetts not now considered the location of significant industrial activity. The importance of early colonial ironworking industries in the everyday life of southern New England residents – farmers making their own tools but also merchants, schoolteachers, and others for whom bog iron bar stock was a form of currency – is one more chapter in the bog iron story waiting to be explored. That this milieu also engendered successful industrialists like Oliver Ames (shovels) or important gunsmiths (Hugh Orr) or cannon forgers (Peter Oliver) illustrates the complexity of the impact of what seems at first to be a simple obscure and now forgotten colonial activity, the harvesting and smelting of bog iron.

Iron Bar Stock as Currency or Commodity

In supplying bar iron to the shipsmiths of the Buzzards Bay region, the forges of the Carver and Wareham area would have had strong competition from the many bog iron blast furnaces and forges located to the north in the Taunton River watershed. This watershed was much larger and was where the Leonards had forges at Two Mile River, which operated almost 70 years before the first furnace at Popes Point went into blast in 1720. Also supplying the early colonial shipwrights of Buzzards Bay, Narragansett Bay, and Newport, RI, were a bloomery at Mill River in Taunton (1670), the Stoney Brook Forge in Old Norton (1698), forges in Bridgewater (Town River, 1707), North Easton (Trout Hole Brook, 1716), and many other forges and furnaces in South Easton, Easton, North Easton, Taunton, and Middleboro that had sprung up on virtually every stream and water privilege by the early and mid-18th century. All of these forges and furnaces were supplying bar iron and malleable iron to blacksmiths for forging locally made horticultural tools or were the locations where bog iron blooms were made into tools. The observation that many hand tools were forged on site by individual families on their farm forges is a recurring theme in the writings about the toolmaking activities of the colonial period (Sloane 1964). We already know that many a farm family earned extra income by harvesting bog iron and bringing it to the forge to be smelted into a bloom of wrought iron. Emery (1893) provides a detailed listing of many examples of the use of iron bar stock, often in units of hundredweight, as a form of species.

> An order from an early settler to pay Mr. Greene the "schoolmaster's rate":
> *Ensine Leonard*, I pray to let M' greene have four shillings more in iron, as money, and place it to my account
>
> JAMES WALKER.
>
> June 20, 1684. (Emery 1893, 623)

Emery's listings of the use of iron as a form of money indicates that this custom, which originated as soon as the Leonards were making iron at Taunton in the mid-1650s, continued into the early 18th century.

The importance of iron as both a commodity and a trade item raises the possibility that many forges sold muck bars to individual farm families and to the small smithies that were located in every town and village. This wrought and/or malleable charcoal iron would have been made into the hand tools of the colonial period by village and farm smiths whose operations were too small to be documented in town histories and who, in almost all cases, did not sign their tools because they were making them for themselves or for their village neighbors.

The blast furnaces of Carver and elsewhere were larger and more well documented commercial operations than the village blacksmith. They were providing pig iron for molders to cast hollowware on site that was then shipped throughout the colonies. Less obvious is the role of these furnaces and forges in supplying ironware for shipsmiths a century prior to the heyday of the New Bedford whaling industry. So many forges and furnaces had sprouted up in the bog iron country of southeastern Massachusetts during this period that it would be highly unlikely that the shipwrights who built hundreds of small fishing vessels and coasting schooners in this region in the 18th century would have needed to import bar iron from England. The same observation may be made about the numerous anchorsmiths whose forges were located throughout bog iron country. The anchors made at these forges were corrosion-resistant wrought iron and were not made from cast iron. The siliceous bog iron of this region was particularly conducive to anchor forging. Numerous anchor forges are noted in town histories, though many others escaped documentation. The fact that large quantities of Swedish bar iron were also imported from England, especially in the early 19th century, usually for special purpose applications such as making blister steel for edge tools, has helped obscure the extent of the vigorous bog iron based industries of southeastern Massachusetts.

To the west of these bog iron deposits, the largest of which were in the Carver and Middleboro townships, there are few, if any, known bog iron deposits of significance in the watersheds of the many rivers of Connecticut, which later became such important tool producing centers in the 19th century. One of the many mysteries is where the pig iron and wrought iron muck bars shipped down the Taunton River and the Wareham Narrows was ultimately fined into other products. Bining (1937) suggests that the blast furnaces of southeastern Massachusetts limited their production to hollowware; at least until 1720 there were few other indigenous sources of cast iron (pigs and sows). After 1720, Pennsylvania, Maryland, New York, and New Jersey produced much more wrought iron bar stock and cast iron than southern New England ironworks. This raises the question: were bloomery produced muck bars (wrought iron bar stock) the main source of iron for southern New England shipsmiths, especially in the early and mid-18th century? The same question may be asked about the iron smelted at the Carver blast furnaces between 1720 and 1820. The workers associated with the Leonard clan, who worked in Raynham and the many other communities where they established forges, grew into a much larger community of ironworkers, who manned the integrated ironworks at Furnace Village in South Easton, Peter Oliver's famed ironworks on the Nemasket River in Middleboro, the ironworks on the Town River in Bridgewater, and the ironworks at Forge Pond in South Canton, where the Kingsley Iron and Machine Company later became the site of the Revere Copper Works. It was in Bridgewater that Thomas Ames first took up blacksmithing around 1730, and his son, Captain James Ames, began making shovels at his Town River forge in 1773, teaching his own son, the famous Oliver Ames, the trade of shovel-making. If any of the pig iron made at the Carver blast furnaces was shipped north to supply the ironworks at Furnace Village in South Easton or the shovel factory at Shovel Shop Pond in North Easton, no record of such activity has yet been uncovered. The locally produced hollowware from this era is well documented. These furnaces also supplied the 19th century cupola furnaces and nail and tack factories of Wareham and Sandwich. The question arises as to how much bog iron country blast-furnace-derived pig iron was refined into wrought and malleable iron or whether most of the latter was produced in local direct process bloomeries. Other questions remain:

- To what extent did New England coasting vessels supply Connecticut shipbuilding communities, such as those at Mystic, with pig and bar iron from watersheds such as the North and Taunton rivers?
- When did supplies of bar iron from Pennsylvania and elsewhere begin to supplant the bar and pig iron from southeastern Massachusetts, the bog iron deposits of which were being rapidly depleted by the late 18th century?

These are, as of now, unanswered questions and will have to await further research by other students of the history of the American iron industry.

When Oliver Ames began making his shovels, first at Bridgewater and then at North Easton after 1803, one of the principal destinations of his shovels was, surprisingly, Newport, RI, where he brought his shovels first with a one horse wagon and then in much larger quantities in wagons pulled by six oxen (Galer 2002). The fact that overland transport by horse and wagon, and later by oxen, was utilized in the flatlands of southeastern Massachusetts to bring iron stock south to Newport and Wareham also raises the question as to what extent ironwares and bar iron were brought north to Boston by the same method from the burgeoning iron and tool producing centers at North Easton, Middleboro, Bridgewater, and those along the North River watershed. Large numbers of fishing vessels, coasters, and East India merchantmen were being built to the north of Buzzards Bay along the shores of Cape Cod Bay and Massachusetts Bay and at the outfalls of the Merrimac and Piscataquis Rivers. Documentation of the trading routes used by the bog iron industries of southeastern Massachusetts would help us determine the extent of the role these industries played in supplying bar iron to the many shipsmiths who worked on the coast to the north.

The finery forges of local blacksmiths working as shipsmiths were where fittings for ironing the whalers of New Bedford and Nantucket were forged. There are no surviving records indicating bog iron harvesting and smelting on Nantucket Island. The live oak, white oak, and yellow pine used for the construction of the Nantucket whaling ships and merchantmen originated from the southern states in a range from Virginia to Texas (Little 1971). Georgia and South Carolina were a particularly important source of yellow pine for planking these ships and many others built to the north of Cape Cod and downeast in Maine. Most of the Nantucket whaling ships were constructed in the shipyards of Old Rochester, MA, which still had access to the huge pine forests and probably to some white oak in inland southeastern Massachusetts, at least in the early and mid-19th century. The Carver blast furnaces and those at Easton, Middleboro, and Bridgewater had access to the largest bog iron deposits along the Atlantic coast north of the pine barrens of New Jersey.

The bog iron industries of southeastern Massachusetts began fading after the end of the War of 1812. The end of that war brought a period of prosperity to America; few if any privateers traveled the New England coast. The days of the Nantucket and Martha's Vineyard privateers hiding at Bunker Hole and spending their summers on Head Harbor Island in downeast Maine had also come to an end. Huge rock ore deposits, especially in Pennsylvania, but also in Connecticut, Maryland, and New York, were already being exploited by the 1730s. Important ironworks had already been established at Tinton Falls, NJ (1682); the famed ironworks at Principio, MD, at the headwaters of the Chesapeake Bay were operational by 1722 (Gordon 1996). Pennsylvania produced bar iron would also have been a principal source of the iron needed by the shipsmiths of New England by the late 18th century. The American coasting fleet was small enough in individual ship size (50 to 75 tons with a draft of six feet or less) that coasting schooners could travel up to the head of the tide of almost every major river drainage in the east coast of the United States. So many hundreds of coasting vessels were built and were plying their coastwise trade by the late 18th century that no generalizations can be made that bar iron from the bog iron country of southeastern Massachusetts was the principal supply of shipsmiths or anchorsmiths anywhere.

In New Bedford, Charles W. Morgan (d. 1861) had become an active participant in the New Bedford whaling industry as a partner of William Rotch, Sr. and Samuel Rodman, Sr. in their merchant shipping business in the 1820s. Later in his career, Morgan went to Pennsylvania and purchased a blast furnace at Bloomsburg and the Duncannon Ironworks to supply his shipbuilders at Old Rochester and elsewhere. His focus on Pennsylvania as a source of the ironwares he needed is indicative of the decline of the Taunton River watershed as a source for the iron bar stock used to make the iron fittings needed on the New Bedford whalers. Other than Oliver Ames' shovel factory, which soon began using steel brought by train, also probably from Pennsylvania, most of the furnaces and forges of the bog iron countryside had closed. But the Ames shovel factory was not the only legacy of this earlier industry. At this point in time, Bridgewater was the 5th largest community in Massachusetts and a relatively large series of foundries and factories had grown up around the Town River, where the Bridgewater foundry supplied the iron sheeting for the USS Monitor. Charles Morgan continued to be an influence on the local iron industries, which still flourished in the form of nail factories in the Wareham area, where its many cupola furnaces and nail factories, including an iron foundry owned by Morgan, were the last best customers of the many farmers and workmen who continued to harvest bog iron after the decline of this resource in the 1830s and 1840s (Doherty 1976).

By this date, as noted elsewhere, radical changes were soon to occur in the industrial landscape of New England and the states to the south. Railroads would soon spread their web everywhere. The role of the coasting schooners and their cousins, the coasting steam packets built in New York City in such great numbers, was about to be greatly diminished. Only the fuel savings provided by the sleek larger three- and four-masted schooners soon to be built in Penobscot Bay to the north, the famed Downeasters, would provide competition in the battle over who was to be the bulk cargo carrier for lumber, cotton, coal, lime, and granite. When railroads spread through New England in the 1840s and 1850s, both the wagon transport of Oliver Ames' shovels and the services of many a coasting vessel became unnecessary or irrelevant. The whaling industry would last for another three decades.

By the beginning of the heyday of whaling ship construction (1820), iron bar stock as a form of currency was a century or more in the past. But as a commodity of importance for the New England shipwright and his enablers, the shipsmiths and edge toolmakers, iron bar stock and its daughter product, steel, were still of critical importance to the many shipbuilding communities of mid-19th century New England. But the era of the bog iron bloomeries and forges of New England as suppliers of essential ironwares was over. Multiple sources of iron and steel were supplying the booming shipyards of New England. The age of wooden shipbuilders and the maritime culture, which was its achievement, flourished and then gradually perished as the American factory system and its railroads began undercutting its hand-hewn forge welded infrastructure. The last chapters in our review of the ferrous metallurgy and edge toolmaking of New England's maritime history narrate selected stories about coastal Maine as America's most important shipbuilding center after 1840.

IV. Maine Shipsmiths and Changing Technologies

1840: The Beginnings of Change

The growth in the shipbuilding industry along the Maine coast in the 19th century occurred during a period of rapid change in the industrial landscape of New England. By 1840 in southeastern Massachusetts, the operation of most blast furnaces that smelted locally mined bog iron had ceased. At the same time, the foundries and forges of southeastern Massachusetts continued operations and, in many cases, expanded their range of manufactured goods. The iron and steel they needed was brought by coasters to inland ports such as Taunton and soon by rail to communities with no access to sailing vessels, such as Easton, Stoughton, and Bridgewater. The diminishing productivity of the resource-based bog iron industry contrasted with the ironic growth of iron and tool producing local industries; these changes were a microcosm of the larger shifts now occurring in the American industrial landscape. In 1840, the spread of the railroad was in its infancy. The introduction of firearms and machinery made with interchangeable parts made by other machines was just beginning. Pig and bar iron and steel were readily available from a multitude of foreign and domestic sources, including England, Germany, Russia, Spain, and Austria. These were supplemented by the vigorous production of the Pine Barrens of New Jersey (bog iron) or rock ores from Salisbury, Connecticut; Pennsylvania; New York; West Virginia; and, later in the century, Minnesota. In 1840, Oliver Ames was transporting his shovels by oxen to Newport, Rhode Island; by 1860, the availability of rail transport allowed a rapid expansion of his business. After 1860, the products of the Collins and Douglas ax companies could be easily shipped to western states, supplying growing agricultural communities in Ohio and elsewhere, which would soon spawn their own tool manufacturing businesses. The products of the shipsmiths who ironed the whalers of New Bedford or the clipper ships of East Boston were being replaced by the products of foundries manufacturing malleable cast iron ship fittings. Edge tools would soon be made from hot rolled cast steel in large factories, such as those operated by the Buck Brothers, Thomas Witherby, or the Underhills, who, by the end of the Civil War, began providing significant competition to smaller edge toolmakers working in local communities, who had previously been the main source of tools used by the shipwright.

The New England shipbuilding industry continued to expand during the period from 1840 to 1856, despite the greatly diminished role of bog iron as a source for iron fittings for ships. Boston and Essex remained important New England shipbuilding communities, while activity declined in Cape Cod Bay and Newburyport, one of New England's most active shipbuilding towns in the late 18th and early 19th century. The last ship built on the

North River in 1871, the *Helen M. Foster*, was built by Joseph Merritt, who is of interest here as a relative of Henry Merritt who built the *Sarah Jane* in 1851 (Briggs 1889, 257). (See the chapter "Lost and Found History".) The largest growth of the shipbuilding industry occurred in Maine, where the nearly invisible industry of eastern Maine (see the Appendix listing the numerous ships built east of Belfast in the National Watercraft Collection) was soon supplanted by the amazing productivity of the Waldoboro Customs District (Thomaston, Warren, Waldoboro, and Boothbay) in the 1840s and early 50s. At this time, Bath, Maine, and other communities along the Kennebec River, long an important shipbuilding region, began their 50 year reign as America's most important shipbuilding center, first challenging, and then superceding, Boston in the tonnage of ships produced. Shipbuilding design would reach a level of perfection in the years after the Civil War in the form of the downeasters of Penobscot Bay and the schooners of Waldoboro, Damariscotta, and Bath, which would be built until the end of the century. Chapelle (1960) notes that these downeasters, built in the years after the depression of 1857 and the Civil War curtailed shipbuilding activities, "represented the highest development of the sailing merchant ship" (Chapelle 1960, 49).

Figure 36 Cast steel slick marked "C. STILLER" "ST. JOHN" and "CAST STEEL WARRANTED", 29 ½" long including the wooden handle, 15 ½" long blade that is 4" wide tapering to 1.5". This slick is unusual; instead of being flat across, it has a central ridge with slightly slanted sides but no beveling. In the collection of the Davistown Museum 030505T1.

In New Bedford, the career of Charles Morgan and his acquisition of Pennsylvania iron foundries in the 1840s and 50s helps document America's changing industrial landscape. Coasting vessels could bring iron bar stock from anywhere. After 1850, the web of railroads spread across America and Maine, quickly replacing the coaster as the most important mode of transport for materials used by America's growing factory systems. Bangor, Maine, in particular, was one of America's busiest seaports in the 1830s and 1840s. While exclusively a lumber port, Bangor was also the birthplace of the innovative toolmaking efforts of several pioneers of the classic period of American toolmaking, such as Edward H. Bailey, Michael Schwartz, and Samuel Darling, the latter of whom later played an important role in the rise of Darling, Brown & Sharpe, later the Brown & Sharpe Company of Rhode Island. Along with many other companies, they used newly redesigned measuring tools, including depth gauges, try squares, and straight edges, to build the machinery that replaced some of the hand tools used in New England's shipyards and forges. The shipbuilding activities along the Maine coast were intimately linked to this timber-harvesting community, which flourished and then faded, foreshadowing the end of the wooden shipbuilding era. The hand tools designed and manufactured by Schwartz & Darling in Bangor in the 1850s are now a forgotten footnote to the gradual switch from hand-built to partially machine-built sailing ships.

Figure 37. From top to bottom: Whorff, Madison, Maine, broad ax, 10 ¾" long, 7 5/8" blade, 3" poll, Davistown Museum 21201T2. Thaxter, Portland, Maine, mast ax, 11 ¾" long and 7" wide blade, 3" long and 1 3/8" wide rectangular poll, Davistown Museum 91303T20. Lovejoy, Chesterville, Maine, broad ax Photo used with permission from Philip McKinney. Bragg broad ax, 9" long, 7 ½" wide head, Davistown Museum 062603T1.

The context shared by all Maine shipsmiths, edge toolmakers, and shipwrights was an expanding economy undergoing industrial development and a rapid switch to steam- and water-powered machinery that replaced handwork with industrial production after 1840. While the shipsmiths and edge toolmakers of Maine were still forge welding their ship fittings and edge tools, toolmakers to the south, such as those at the Collins Axe Company, began using machinery to make their drop-forged tools. The evolution from the handmade ax to the all cast steel ax is well documented (Kaufman 1972, Heavrin 1998). The irony of the activities of both the shipsmiths and the edge toolmakers of Maine is their conservatism and reliance on traditional methods and techniques, in the context of the onslaught of an industrial age that would eventually, but with great difficulty, render the sleek schooners and downeasters they were now building obsolete. The mystery remains unsolved: where did the edge toolmakers of Maine obtain both the charcoal iron and the steel for the tools that they made? Did they obtain iron bar stock from Pennsylvania and Maryland as ballast in coasters and later by rail? Was Sweden or Spain, via the transatlantic trade, a source? Were any blister steel furnaces operating in Maine, or did Maine edge toolmakers import domestic blister steel from New Hampshire or southern New England forges, before perfecting the art of making shear steel? Did they import any crucible steel from England, and when did it start coming from Pittsburg, where Singer, Nimick, and Company began making high quality cast steel in 1853 (Barraclough 1984a, 148)?

The Katahdin Ironworks began operation in northwestern Maine in 1843 but had no known role in supplying iron to

Maine's shipyards, shipsmiths, and edge toolmakers. Yet the most important primary historical evidence, the tools themselves, present in the form of signed, forge welded edge tools, illustrate the fact that the mid-century florescence of shipbuilding in Maine was accompanied by a vibrant local edge toolmaking industry. The listings in the 1856 Business Directory (Parks 1857) further illustrate the extent of this indigenous industry. Some tools were also imported from the growing factories of New Hampshire, Massachusetts, and Connecticut. The importance of the edge toolmakers in the St. Stephen and St. John area of Canada along the northern shores of the Gulf of Maine is less well known. These edge toolmakers also helped supply Maine's shipwrights and timber harvesters with edge tools. Josiah Fowler, of St. John, NB, and the Broad clan of ax makers at St. Stephen, NB, are still remembered as among the most important sources of edge tools used in New England (Gardner 1975, Klenman 1990).

These Canadian toolmakers and many edge toolmakers listed in the Maine Business Directories (Parks 1857) and the *Registry of Maine Toolmakers* (Brack 2007) did not have access to significant quantities of American-made cast steel before 1865, which was only available from Singer and Nimick in the 1850s, and widely distributed only after the end of the Civil War. It is not known whether any Maine edge toolmakers or more famous toolmakers, such as Thomas Witherby or the Buck Brothers, purchased this domestically-made cast steel, made their own cast steel, or imported English cast steel in the years before Pittsburg began competing with Sheffield for this market. Only occasionally are edge tools made before 1850 found with the mark "cast steel," suggesting that English cast steel was imported by Maine edge toolmakers. During the boom years of Maine's shipbuilding (1830 – 1857), Maine's edge toolmakers forge welded tools of the highest quality from iron and steel available from a wide variety of sources. The tools themselves linger on, "accidental durable remnants" of the skilled artisans of the early and mid-19th century. The secrets of this ferrous metallurgy are buried with the edge toolmakers who forge welded their usually gleaming surfaces.

Locally produced bog iron may still have been used by downeast shipsmiths to make edge tools, as illustrated by some of the surviving tools made by the Ricker family of Cherryfield. Bog iron was probably readily available in downeast Maine in the blueberry barrens above Cherryfield and Harrington. But where did the Billings clan and the many other ax makers of Oakland and Clinton, ME, for example, located well up the Kennebec River system, get their iron and steel? The same question can be asked about the many other edge toolmakers working in the vast Kennebec River drainage and in the lower Penobscot River area, making not only adzes and chisels, but the ubiquitous New England pattern broad ax, which is much more common in New England tool chests and collections than the larger Pennsylvania pattern broad ax. A common variation on the New England pattern broad ax is the slightly trimmer mast ax that was also an essential tool in Maine and New England shipyards. Did any Maine toolmakers pile and reforge blister steel bar stock into the high quality soft but durable steel needed for edge tools? Or did Maine edge toolmakers import tool steel along with bar iron from Pennsylvania, New York, or West Virginia, areas that had entirely replaced southeastern Massachusetts as a source of iron and steel used in hand tool production by the early 19th century?

New England toolmakers and those in other states always marked their tools with the maker's name and often with the place of manufacture. These marks constitute clear evidence that these tools were not made in England and imported to the United States. English-made edge tools were even more clearly marked and may have been more sophisticated looking as a result of their artful forging by the experienced edge toolmakers of Sheffield. No New England style broad ax is known to the author to have a Sheffield touchmark. By the 1840s, if not earlier, edge tool marketing was a regional affair, with Whorf or Thaxter broad axes being used not only in local communities but also surfacing in any shipbuilding community in the region. A New England pattern trimming ax signed A. LOVEJOY CAST STEEL is illustrated in Heavrin (1998, 138), and many other edge tools made by the Lovejoy clan from the Chesterfield, ME, area have surfaced in collections and used tool stores (e.g. Liberty Tool Company). Some are marked cast steel, and this is not surprising since the Lovejoy clan were still actively making axes into the late 1870s and American cast steel became readily available and accessible by train after 1865. We can at least suspect that Lovejoy broad axes that are not marked cast steel were probably made before 1865; this observation would also apply to many other New-England-made edge tools if they were not made of imported English cast steel. Whenever English cast steel was imported and incorporated in American made edge tools, the superiority of English cast steel would usually be advertised as a selling point by the ubiquitous trademark "cast steel" or "warranted cast steel." Large numbers of high quality edge tools made before the Civil War not so marked indicate the presence of a robust indigenous edge tool manufacturing industry in the early 19th century. This industry clearly has roots in the flourishing ironworking industry of colonial era New England. It is probable that a significant portion of the best quality edge tools not touchmarked "cast steel" were made from repiled and reforged blister steel, i.e. shear steel. The wooden ships built in New England's 19th century shipyards, including those along the Maine coast, have long since rotted into oblivion. Many iron and steel tools, having the benefit of the shelter of the tool chest or workshop, have survived to help us unravel part of the tale of the preeminence of both New England shipbuilding and toolmaking in the 19th century. We still do not know how and where much of the steel incorporated in these welded iron-steel edge tools was made.

The period between 1840 and 1856 is characterized by rapid technological change even in the conservative shipbuilding communities of coastal Maine. Most Maine shipwrights clung to traditional methods, tools, and techniques for building wooden ships. When Maine business directories and registries began appearing in the mid-19th century, the anonymous blacksmiths of early colonial coastal Maine, who were the shipsmiths for locally built shallops and brigs, became a specific category of artisans who, along with edge toolmakers and blacksmiths, warranted their own separate listings in these business directories. The listings of shipsmiths in these directories are, along with the survival of the edge tools that they forged, the principal evidence of their presence in the 19th century. These registries and directories and the half models in the *National Watercraft Collection* (Chapelle 1960) remind us that Bath was not the only center of shipbuilding in 19th century Maine. Baker (1973) has left us with an invaluable detailed description of the shipbuilding industry of the Bath-Kennebec River region. Eaton (1865), Packard (1950), and Stahl (1956), among others, have made major contributions to documenting shipbuilding along the central Maine coast. Ironically, none of these writers mention shipsmiths or explore the issue of how and where the tools used to construct the ships they are writing about are made. Other than as noted above, shipbuilding activities east of the Penobscot Bay are nearly undocumented. More well documented are the numerous shipyards of the Waldoboro Customs District including Thomaston, Warren, and Boothbay. Obscured by the fame of the ships built in these communities and the shipyards in which their construction occurred is the identity of the shipsmiths and edge toolmakers that played such an important role in their construction.

The Shipsmiths of Thomaston, Warren, Boothbay, and Waldoboro

Figure 38 Vaughn & Pardoe gutter adz, forged and weld steel with a 16" wooden handle, 10 ¾" long and 2 ½" diameter adz. A welded steel-iron interface is clearly visible on this tool, which also shows extensive evidence of forge welding. In the collection of the Davistown Museum 61204T2.

The success of the shipbuilding industry of the central Maine coast in the 1840s and 1850s has its roots in a long tradition of shipbuilding in New England. Early colonial shipbuilding activities in Maine east of Kittery are nearly undocumented, but the thousands of settlers who moved to coastal Maine between 1623 and the diaspora of the Indian Wars in 1676, which cleared the coast of Maine of any settlements east of Wells, must have built hundreds of shallops, small fishing vessels, work boats, and brigs. These small vessels were essential for the survival of Maine's settlements, which flourished as fishing communities and participants in the coasting trade. By 1650, a firewood- and hay-starved Boston was already dependent on Maine's coastal resources. This dependence, in turn, engendered the participation of Maine's new settlers in fishing, coasting, timber harvesting, and small vessel construction activities. Voyages to and from the markets at Boston, Salem, and Portsmouth were the primary means of supporting and supplying the residents of the Province of Maine. Slyvanus Davis of Wiscasset is one example of a trader and entrepreneur already active two centuries before the boom years of the Waldoboro Custom District. Maine's earliest know blacksmith, James Phipps, father of the Royal Governor, Sir William Phipps, was already working at Pemaquid by 1625 and was almost certainly a shipsmith. The Clarke and Lake trading station at Georgetown Island was also the location of a mid-17th century ironworks. Shipsmithing was probably the most important of all smithing activities at this early settlement.

The tools and technologies used by Maine artisans for timber harvesting and ship construction remained essentially unchanged for the next two centuries, despite the evolution of new steelmaking strategies, which characterize this period. The shipwrights of the Waldoboro Customs District had, in fact, the legacy of two centuries of shipbuilding experience by colonial and early American artisans to provide the basic knowledge of "how to" or "making do." When the white oak forests, whose range extended from south central coastal Maine to Georgia, were quickly depleted in New England by the late 18th century, hackmatack was used for ship's knees in situations where it was inconvenient to import white oak from the forests of the Carolina coastal plain. It was the broad ax that was used to cut and shape (beat out) ships' knees; we know more about where the lumber was obtained for Maine's shipwrights than where the steel was produced that Maine edge toolmakers used to forge the tools that cut and shaped the white oak, hackmatack, yellow pine, oak, spruce, and other woods used in ship construction.

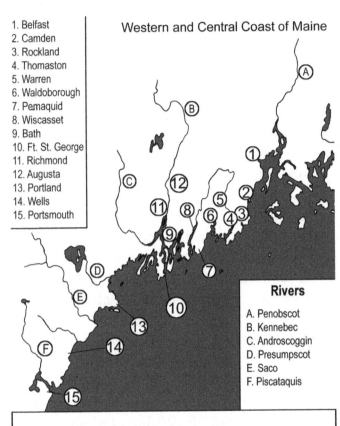

1. Belfast
2. Camden
3. Rockland
4. Thomaston
5. Warren
6. Waldoborough
7. Pemaquid
8. Wiscasset
9. Bath
10. Ft. St. George
11. Richmond
12. Augusta
13. Portland
14. Wells
15. Portsmouth

Western and Central Coast of Maine

Rivers

A. Penobscot
B. Kennebec
C. Androscoggin
D. Presumpscot
E. Saco
F. Piscataquis

Figure 39 Western and central coast of Maine.

The shipsmiths of the Waldoboro Customs District, many living downstream from the water mills of the Davistown Plantation on the Georges, Medomak, and Sheepscot rivers, had little access to local bog iron deposits. Iron and steel for ships' fittings and edge toolmaking had to be brought to Maine from other locations. Sweden had long been a major supplier of the best charcoal iron for edge toolmakers. The high silicon content of wrought iron made bog iron ideal for the anchorsmiths of southern New England and for the manufacture of the harpoons made by the New Bedford whalecrafters. Before the appearance of steel furnaces in the colonies, the manganese content of hydrated bog ore may have assisted limited edge tool production in the colonies, but wrought iron with a high silicon slag content was not the form of smelted iron most valued by edge toolmakers. Charcoal iron, carefully refined and in many cases double-refined from wrought iron to reduce its silicon slag content was the probable choice of forge masters making blister steel out of bar iron. The unanswered question is: when did high quality charcoal iron produced in Pennsylvania and upstate New York begin to supplant bar iron imported from Sweden, Russia, and elsewhere for use in America's blister steel furnaces? In the flourishing shipbuilding towns along the central Maine coast including those in the Waldoboro Customs District, where did shipsmith and edge toolmakers get their iron and steel and how were their edge tools forged?

124

At the same time that America's industrial landscape was sprouting roots in locations with access to water power throughout New England, the shipbuilding industry in Maine was experiencing amazing growth. Maine Business Directories and Registries help us track many of the participants in the florescence of Maine's shipbuilding industry. Ironically, no shipsmiths are listed in the Maine Directory for 1856 in Bath, but Bath shipsmiths suddenly reappear in the directories of 1881 and 1882. Numerous edge toolmakers who made the tools for the shipwrights of Maine are listed in the directories throughout the period. By 1840, Maine was producing more ships than any other state except Massachusetts and would soon eclipse that state by the end of the decade. One major center of shipbuilding activity in Maine was the Penobscot Bay region from Belfast southwest to Boothbay and Wiscasset, the Waldoboro Custom District. Now nearly forgotten, downeast Maine east of Penobscot Bay was a second important shipbuilding region. The third area of activity, soon to become America's most important center of shipbuilding, included the Kennebec tidewater communities of Bath, Dresden, Richmond, and Bowdoinham.

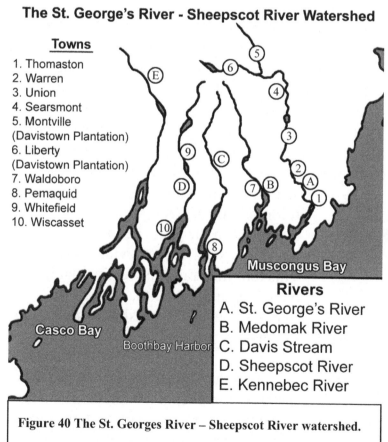

Figure 40 The St. Georges River – Sheepscot River watershed.

In the 1840s, the vigorous productivity of the Waldoboro Customs District, which included Thomaston; Warren; and Boothbay, marks a unique time in America's industrial history. During this decade, the small communities within the Waldoboro Customs District were producing as much as 8% of the wooden ships built in America. They were supplied both with tools and wooden products by upstream forges and mills on the Georges, Medomak, and Sheepscot rivers (Fig. 40). The American factory system was late to arrive at these shipyards; no steam-powered tilting band saws or massive planers or table saws were used in these yards at this time. The work of the shipwright was still entirely dependent on traditional hand tools, such as pit and frame saws, New England pattern broad axes, Yankee pattern lipped adzes, augers, mast shaves, slicks (slices), spar planes, caulking irons and mallets, and the ubiquitous New England felling ax. Many of the toolmakers supplying these tools still lived in or near these shipbuilding centers or in upstream locations with sufficient water power for trip hammer operation.

Figure 41 Gouge, cast steel and wood, 14" long with 4 ½" handle, 1 1/4'" wide cutting edge, signed "Vaughan Pardoe & Co Warranted Union." In the Davistown Museum collection 111001T3.

During this period, Vaughn and Pardoe made edge tools in Union, ME, (Fig. 41) located near Waldoboro, and they were probably also shipsmiths, supplying fittings for nearby shipyards. James Mallet made edge tools at Warren (Fig. 8) and John Mallet, probably his brother, made edge tools and ships' fittings at nearby Rockland (then part of Thomaston). Benjamin Kelley (Fig. 43) was beginning his career as a prolific edge toolmaker working in Belfast. All are represented by tools in the collection of The Davistown Museum and were toolmakers during the heyday of shipbuilding in the Waldoboro Customs District.

Just to the south of Warren and Thomaston on the Sheepscot River at Whitefield, Peter King (d. 1858) achieved wide renown for the quality of his edge tools as illustrated by the elegant mast ax shown in Fig. 42. In Portland, Higgins & Libby were equally famous for the quality of their edge tools. The mortising gouge (Fig. 42 and on the cover), formerly on exhibit at the Maine Historical Society and now part of The Davistown Museum exhibition "The Art of the Edge Tool," rates among the finest edge tools ever recovered from New England tool collections by the Liberty Tool Co. No imported English edge tool surpasses this carefully forged implement, which, because it is not labeled "cast steel," can be assumed to be made from reforged blister steel (shear steel) fashioned in the years before American-made cast steel became widely available. This tool illustrates the reality that in New England, rather than having to import expensive English cast steel, edge toolmakers were capable of taking the products of American steel furnaces and reforging them into the finest edge tools, just as German edge toolmakers had done in the Derwent River valley of northeast England at the beginning of the 18th century before English cast steel became widely available.

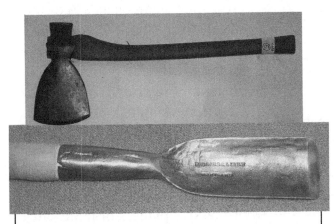

Figure 42 Top: King mast ax from the collection of Roger Majorowicz. Middle: Higgins & Libby mortising gouge from the Davistown Museum collection 61204T17.

The heyday of the Waldoboro Customs District was brief: The town of Warren, located at the head of tide on the Georges River just above Thomaston, built no ships after 1863. Thomaston, Waldoboro, and to a lesser extent, Boothbay, continued as important shipbuilding communities well into the late 19th century, but never again produced such a large percentage of American-built ships as they did in the 1840s and 1850s. Perhaps the greatest contribution of these communities was the high quality and exquisite designs of the ships built in the post-Civil War era. Just one example of the finesse of the Maine shipwrights is another icon of New England's maritime era, the *Flying Cloud* of Damariscotta built in 1851. By this date, power tools may have aided the construction of these ships but the final product was the result of decades of design, layout, and framing up experience combined with the expertise in the use of the edge tools of the shipwright that comes with centuries of experience.

Bath continued to build four-, five-, and six-masted schooners, each weighing thousands of tons, until and after the end of the century. The famous downeasters, sleek bulk cargo carriers built in Penobscot Bay between 1860 and 1885 in towns such as Belfast and Thomaston, were another example of a specific form of sailing ship, i.e. bulk cargo carriers, that represented, with the schooner, a final mastery of the art of building a wooden ship.

After 1865, steam-powered tools became an increasingly important component of the larger shipyards, relieving the labor of much of the most tedious hand work of the shipwright. Labor, and thus crew-saving, steam-powered winches and windlasses began making their appearance after 1870, allowing continued competition between the massive bulk cargo schooners and coal-fired ocean going steamers in the race to industrialization that resulted in the end of the era of the wooden ship.

The vigorous community of edge toolmakers, shipsmiths, and shipwrights

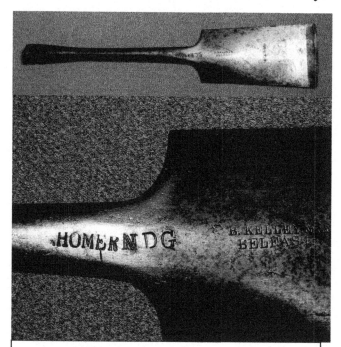

Figure 43 Ben Benjamin Kelley slick and detail, forged or cast steel, 16 ½" long, 6 7/8" long body with a 3 11/16" wide blade, signed "B. KELLEY & CO" "BELFAST" with owner's mark "HOMER N D G" in the Davistown Museum collection 040904T1. This tool appears to be all steel, with a higher quality cutting edge welded to the steel body. It is from the Spear Estate, Warren, Maine

127

that evolved in Maine in the 19th century had its roots in 200 years of shipbuilding and iron-smelting in southern New England. In the Province of Maine, the early florescence of shipbuilding and some early colonial iron-smelting and toolmaking activities were interrupted by the French and Indian Wars (1676 – 1759). This warfare resulted in the abandonment of much of the coast of Maine, while southern New England shipbuilding communities west of Wells continued to grow and flourish. After 1720, isolated shipbuilding communities east of Wells slowly reestablished themselves, and, after the Treaty of Paris in 1763 and the end of the Indian Wars, a vast migration of settlers occupied coastal Maine and its major river basins. This repopulation of coastal Maine was accompanied by a resurgence of shipbuilding and undocumented toolmaking activities that were a prelude to the great climax of shipbuilding in Maine in the 19th century. In both the early colonial period and the post-war migration to Maine, the activities of the shipsmiths were a most important, but nearly forgotten, footnote to the much larger story of the florescence of American toolmaking in the 19th century, which shortly followed.

Bath and the Invisible Shipsmith

From the U. S. Report on Commerce and Navigation came the following tonnages of new vessels built in various cities in 1854:

City	Tonnage
Boston	69,550
New York	63,496
Bath	58,451
Waldoboro	31,476
Philadelphia	24,128

Figure 44 Ship tonnages. Table 2 from William A. Baker, 1973, *A Maritime History of Bath, Maine and the Kennebec Region,* 2 vols, Bath, ME: Maritime Research Society of Bath, pg. 467. It should be noted that the above figures for Bath and Waldoboro include shipbuilding in the communities surrounding these two towns, which were part of their customs districts. Many of the ships built in the Waldoboro Customs District were built in the communities of Thomaston, Warren, and Boothbay. Upriver communities, such as Richmond and Dresden, are included in the Bath Customs District.

Figure 47 T. C. Jackson, Bath, Maine, peen adz, 10" long, 5" wide cutting edge, 2 ¾" peen. In the Davistown Museum collection 52403T1.

Figure 46 J. F. Ames, Richmond, Maine, peen adz, 10" long, 4 1/8" wide adz head, 2 ¼" long, ¾" diameter beveled peen. In the Davistown Museum collection 020807T1.

Figure 45 J. P. Billings, Clinton, Maine, hewing ax, forged iron and weld steel, wood handle, wood and leather blade cover, 4 1/8" long, 6" wide blade, 2 ¾" pole. In the Davistown Museum collection 42604T5

By the middle of the 19th century, Maine was the location of the most productive shipbuilding industry in the world. In 1854, the "city" of Waldoboro had built 31,476 tons of wooden ships, over 7,000 tons more than Philadelphia. Only Boston, followed by New York built more ships than the Bath and the Waldoboro custom districts, the latter of which included Thomaston, Boothbay, and Warren.

Numerous noted Maine edge toolmakers were working just outside of the Waldoboro Customs District, supplying the edge tools needed in the huge shipyards of Bath on the Kennebec River. There are many examples of edge tools from this period in the collection of The Davistown Museum and the exhibition "Art of the Edge Tool." Fig. 46 illustrates a peen adz made by J. F. Ames of Richmond, who is listed in the 1855 Maine Registry and whose family produced edge tools until the end of the century. T. C. Jackson (Fig. 47) was a prolific edge toolmaker working in Bath. Numerous members of the Billings clan made edge tools of all kinds to the north of Bath in the Waterville and Clinton areas on or near the Kennebec River. The Billings clapboard slick and timber-framing chisel in the museum collection probably date from the period between 1840 and 1860 and were almost certainly made for the burgeoning shipbuilding industry located downstream in the Bath area. The clapboard slick, in particular, is an example of a tool that was used to produce an important cargo, i.e. clapboards, transported from Maine by many a coasting schooner launched down the ways of a Maine shipyard. Clapboards may also have been used by a joiner doing finishing work on the cabin of a schooner or brig. The Emerson Stream in Oakland was the location of numerous famed ax companies, which operated throughout the 19th century (Klenman 1990, Kauffman 1972). Many of these companies would also have made edge tools for the shipwrights downriver at the same time that they made axes for the upstream timber harvesters that supplied these shipyards. The Yankee lipped adz made by Billings at Clinton and also included in the exhibition "The Art of the Edge Tool" (see back cover illustration) dates from a decade after the heyday of the Waldoboro Customs District and was likely produced for the shipwrights who continued to work in the Bath shipyards until the end of the century.

Figure 48 Clapboard slick, cast steel, wood handle, 21 ½" long, 4 ¾" wide cutting edge, 14 ½" beveled steel handle, 4 ¾" long turned wood handle, signed "BILLINGS AUGUSTA." In the Davistown Museum collection 52403T3.

By 1858, the Buck Brothers, Thomas Witherby, and the Underhills were making sophisticated rolled cast steel timber-framing chisels and slicks to the south of the shipyards of coastal Maine. During the time between the 1856 and 1882 Maine Business Directories, many smaller shipyards continued the use of hand tools without the assistance of the steam-powered band saw. By 1882, all the larger shipyards still operating in Maine, including those at Bath, had switched to power tools to construct what were, as huge five-, six-, and seven-masted schooners, essentially cotton mills, granite quarries, coal mines, and other industries under sail. Only a few coasting schooners were still being produced after this date, which, ironically, is the era remembered for its perfection of the art of wooden ship design and construction. The vision, or perhaps the specter, of these huge creations, all of which had only a very brief lifespan, helps obscure the creative output of Maine and New England shipyards, which had their roots in the great migration to New England and the necessity of building and ironing ships to survive in the new-found-lands with their vast resources of fish, timberlands, bog iron, and water power waiting to be exploited.

The steam engine, already widely used for water transportation for thirty years, was late in coming to water-powered America, but it eventually arrived. Baker notes the case of George Moulton, who went into business as a blacksmith in 1839.

> In 1842 with steam coming into more general use he moved to a shop on the west side of Commercial Street to build steam engines and boilers with John H. Williams as a partner. The first boiler built in Bath was said to have come from this shop. (Baker 1973, 435)

The railroad had come to Waterville in 1849. The "palmy days" (Baker 1973) of shuttling visitors by steamboat from Augusta to Waterville had ended, as had much of the shipping business from the port town of Hallowell, located just south of the head of tide at Augusta. The introduction of the steam-driven bevel saw in Bath shipyards was only 14 years in the future (Baker 1973, 790). It was an innovation that eliminated much of the hand work of the shipwright. Steam-powered machinery had already arrived in Bath by 1821 in the form of saw mills, which began supplementing the wood cut in the region's many tidal mills, and the wood cut by hand at many Maine shipyards with the traditional tools of the sawyer, i.e. frame saws (the first pit saw), whipsaws, and modern (circa 1750-1800) pit saws. The shipwrights' hand was being replaced by power-driven machinery, though not for all shipbuilding functions. Shipwrights continued to use frame and pit saws for work that still could not be done with power tools. "Long heavy timbers that were required to be curved – keelsons, wales and the like – were whipsawed over a pit until late in the nineteenth century" (Baker 1973, 342).

In Bath, in 1856, the shipsmiths with their forged iron fittings were already being replaced by the foundry, which could make not only cast iron fittings but also malleable iron and iron alloy castings of many varieties. Edge toolmaking had already become a specialized trade, as illustrated by the listings in the 1856 Maine business directory (Fig. 18). Many a shipsmith still lingered in downeast Maine, which had not yet experienced the full impact of the availability of steam-powered machinery. The shipsmith was now becoming a specialized ironworker; cast iron fittings and cast malleable iron replaced much of the forged-iron ship's hardware of the past. The word "shipsmith" does not appear in the index of Baker's (1973) comprehensive survey of Maine's largest shipbuilding community. Only on page 434 does Baker note Job Chapman as advertising as a "ship smith and commercial blacksmith." But that doesn't mean there were no shipsmiths working in Bath or Waldoboro at this time; often they were simply called blacksmiths, as Baker notes while describing the shipyard of Lamont and Robinson, Bath's southernmost shipyard.

> Alfred Lemont of this partnership, born in Bath in 1808, learned the blacksmith's trade as a young man and took contracts to "iron" Bath-built vessels by the ton. Eventually becoming a shipbuilder of note… (Baker 1973, 429)

Shipsmiths continued working throughout the age of the wooden ship; they simply became another anonymous artisan in a rapidly expanding industrial economy. The important role they had played in the growth of the colonial economy had already been forgotten.

Baker (1973) provides insight into the identity of a number of ironmongers and foundries where Bath blacksmith-shipsmiths obtained their iron, including the J.H. Allen Company,

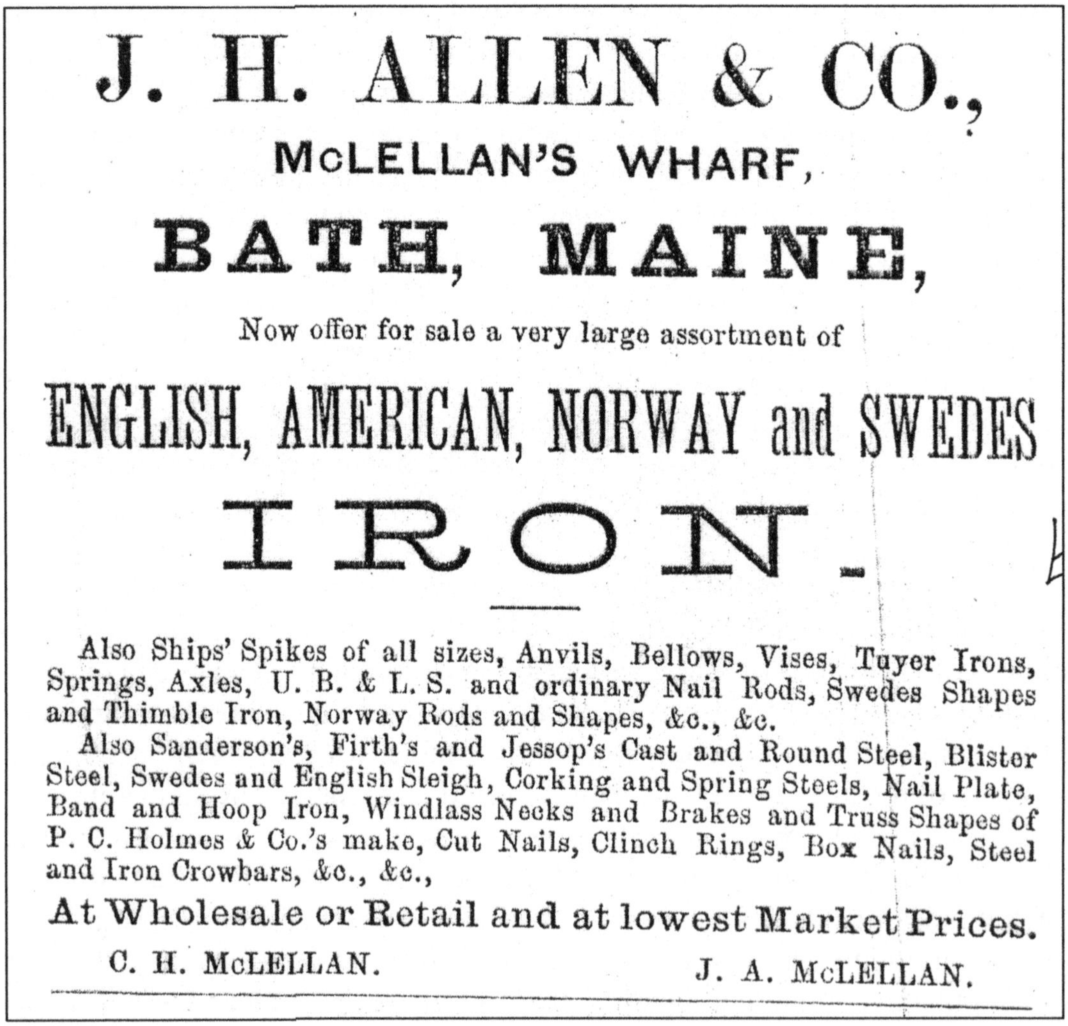

J. H. ALLEN & CO.,
McLELLAN'S WHARF,
BATH, MAINE,
Now offer for sale a very large assortment of
ENGLISH, AMERICAN, NORWAY and SWEDES IRON.

Also Ships' Spikes of all sizes, Anvils, Bellows, Vises, Tuyer Irons, Springs, Axles, U. B. & L. S. and ordinary Nail Rods, Swedes Shapes and Thimble Iron, Norway Rods and Shapes, &c., &c.
Also Sanderson's, Firth's and Jessop's Cast and Round Steel, Blister Steel, Swedes and English Sleigh, Corking and Spring Steels, Nail Plate, Band and Hoop Iron, Windlass Necks and Brakes and Truss Shapes of P. C. Holmes & Co.'s make, Cut Nails, Clinch Rings, Box Nails, Steel and Iron Crowbars, &c., &c.,
At Wholesale or Retail and at lowest Market Prices.

C. H. McLELLAN.   J. A. McLELLAN.

Figure 49 Advertisement.

132

whose advertisement is reproduced in Fig. 49.

> Chapman and the blacksmiths in the other yard probably obtained much of their raw material from J. H. Allen & Company who were Bath's principal suppliers of bar and hoop iron, nails, ship spikes and the like. This firm was located on McClellan's Wharf near the foot of Linden Street. (Baker 1973, 434)

This riverside location suggests that Allen & Company was a wholesaler who probably obtained most of this ironware from incoming ships, and, in turn, may have obtained much of it from Pennsylvania's flourishing forges via the Delaware River and Chesapeake Bay. Liverpool and the Lancashire countryside, as well as Birmingham and the nearby Midlands of the Severn River watershed, were also possible sources of the ironware the Allen Company was selling, but domestic production of hoop iron, nails, and spikes was already well-established in southern New England locations, such as the nail-making factories of West Wareham on the Weweantic River. Pennsylvania was already North America's most productive iron-smelting and -forging area by the mid-18th century. One hundred years after the evolution of Pennsylvania's famous iron industries, Bath and other Maine shipyards thus had the luxury of having access to multiple sources of iron and steel, combined with the ease of inexpensive and efficient delivery services by a long established coasting trade. All would soon change with the coming of the railroad.

Baker (1973) then notes another of the Bath industries that was helping to supply the ironware needed for Bath's many shipyards.

> One of the most flourishing industries in the Bath of 1854 was the iron foundry of C. A. Lambard and Company... its work was confined mainly to the manufacturing of ships' castings the use of which in recent years had grown enormously as iron replaced wood for many items on shipboard. These included "plates for hatches, topmast head caps, frames for ports, belaying pins, pumps, sheaves, capstands, cabooses, etc." (Baker 1973, 434)

Baker (1973) spends eleven pages detailing the supplies and equipment needed to build, equip, and outfit the relatively small schooner the *Adriatic*, launched in 1850, and he notes the many sources of iron fittings used to construct this ship. Much of its ironwork was made locally, but its anchor and chains were purchased from Henry Wood and Company of Liverpool. Also imported from Liverpool were "chain slings for fore and main yards and various sizes of ordinary chain for topsail sheets and runners, topgallant runners and ties, and spencer – and spanker ties – gaff ties, runners, and sheets" (Baker 1973, 446). The James T. Pattern Company of Bath also imported "common English iron, refined flat iron, refined iron, Old Sable iron, and thimble iron" (Baker 1973, 446) from

England. But much greater quantities of iron fittings were made locally by the foundry of William V. and Oliver Moses. "…a variety of cast iron items. Included in the list were 164 pounds of stanchion rings; 1,136 pounds of side ports; 552 pounds of side rings; 375 pounds of hawse pipes; 326 pounds of deck irons; 164 pounds of chocks; 1 ½ pounds of pump standards; 28 shocks for yards; 86 quarter blocks; 44 bushings for change sheets; 64 davit steps; and 84 belaying pins" (Baker 1973, 446). With the exception of the davit steps, all were used for construction of a single 700 ton schooner, the *Adriatic*.

Baker's detailed listing of the wide variety of ironwares used as ships' fittings and hardware and the diversity of their sources illustrates an ironic anomaly of 19th century industrial America. By 1850, Sheffield and nearby Midland areas of Britain had emerged as the world's largest and most productive iron- and steel-producing region. The evolution of Sheffield and Birmingham as England's principal tool- and steelmaking centers helps explain the diversity of sources of hardware available to Bath shipwrights and shipsmiths. English industrial sources thus provided stiff competition for domestically produced iron tools and fittings used to build and iron Bath ships.

When the first English industrial revolutionaries deduced the trick of using fire to do work by heating water, the classic period of the Industrial Revolution they engendered by their inventions had the ironic effect of increasing American dependence on English ironwares just at the moment that political independence from the mother country was achieved. By 1775, America was producing one seventh of the world's iron. The robust indigenous iron-smelting and toolmaking culture suddenly set adrift in its infancy by the English Revolution of 1642 had grown into a pimpled adolescence, with blast furnaces emerging everywhere. (1775). Watt's steam engines and Henry Cort's clever redesign of the reverbatory furnace and his rolling mills came too late to win the Revolution. England's glowing industrial landscape, i.e. little water power and many coke-fired steam engines, was now producing iron and steel in such great quantities that, for the 77 years between 1789 and 1856, it was often cheaper for some American shipyards to import anchors, capstans, windlasses, and other equipment than to manufacture them domestically. Baker's (1973) detailed descriptions of the multiple sources of ironware clearly illustrate this mid-century quandary as does the Sanderson price list of imported and domestically produced steel (Tweedle 1983).

The impact of shipbuilding activities at Bath, Waldoboro, and other Maine coastal towns was to create a huge local and regional demand for goods similar to those listed by Baker and needed for all the other ships being constructed in the heyday of Maine's shipbuilding industry. In addition, with many to be sold in England, Europe, or the southern states once the ships were launched, each needed a cargo as part of its final departure. For Maine, timber, other wooden products (especially staves, clapboards, and house frames), and fish were the prime exports. The prosperity that resulted made inland communities such as Union, Appleton, Searsmont, Liberty, and Montville in the hill country north of Waldoboro and west of Thomaston, flourishing mill towns, at least for the decades in the early and mid-19th century (1820 – 1870) when Maine's shipyards were at their busiest. The prosperity of these communities in the 1840s and 1850s was a result of the economic impact of a local shipbuilding industry, which had begun flourishing two hundred years earlier in Boston and Salem. There was virtually no abatement in the construction of wooden ships for over two centuries in New England, other than the changing regional participation of one city, town, or another in an industry that was the essential ingredient in the success of the American Revolution.

Another fifty years remained for shipbuilding in Maine. The finest designs of the Penobscot Bay downeasters, the sleek schooners from the Waldoboro custom district, and the huge five-masted bulk cargo carriers of Bath's famed shipyards were yet to be built. These mostly machine-made ships were the final chapter in the ironic tale of bulk cargo carriers, which helped facilitate an Industrial Revolution where machine-made hand tools would build the machines that replaced the function of the hand tool. The adz, broad ax, and the slick were the last to go.

1856: Changing Technologies

Shipsmithing doesn't end in 1856, but the date is significant because of the rapid changes in the tools and technologies used by both the shipwright and the shipsmith. Steam engines were beginning to power factories of all kinds throughout North America, allowing their establishment at locations away from water power sources, which had long been the prime movers of America's industrial facilities. In Maine and New England, steam-powered saw mills were being widely established near shipbuilding communities such as Bath, Maine, located downriver from important inland timberlands, replacing those that had traditionally been located on rivers or tidal estuaries. Cast iron power tools of all kinds (table saws, band saws, shapers, drill presses, and other equipment) were being introduced in shipyards in New England and elsewhere. This power equipment paved the way for the construction of the much larger three-, four-, and five-masted schooners of the last half of the 19th century.

Other events marked 1856 as a significant year in America's industrial history. The great depression of 1857 marks 1856 as the last year of a sustained economic growth before the dark years of the American Civil War concentrated much of America's growing industrial capacity on the production of arms. Shipbuilding in New England was greatly curtailed until after the Civil War and, in fact, never returned to its pre-1857 levels. Also of interest was the invention of gasoline at Watertown, MA, in 1856, an event which, along with the concurrent purification of kerosene, soon impacted the viability of the whale fishery, then in its peak years of productivity.

The revolution in the use of steam-powered equipment can be traced to the mid-18th century and the creative designs of the English industrial revolutionaries, a tale recounted in the first volume of the *Hand Tools in History* series. After the English industrial entrepreneurs designed and built the machines needed to make blocks for the English Navy, for example, their unique machines were quickly adapted for the American factory system by American entrepreneurs like Eli Whitney and Simon North. The year 1837 marks the beginning of the use of drop-forged toolmaking techniques at Samuel Collin's ax factory in Connecticut (Muir 2000). Even more important for the New England shipsmith was the reintroduction of the ancient technique of making malleable cast iron from annealed white cast iron, which had been done by the Chinese early in the first millennium before Christ (Barraclough 1984a). Shipsmiths everywhere had traditionally made most fittings out of wrought iron bar stock, sometimes smelted from bog iron by bloomsmiths whose water-powered forges were usually close to the shipbuilding communities they served. Called muck bars in the 19th century, the bar stock that they produced was transported to community blacksmiths and shipsmiths and reprocessed into tools and fittings in bellows-driven forges that have long been forgotten. In colonial times, many of the edge tools that shipwrights used were also hand-forged at such locations, sometimes out of locally produced bog iron and steel bar stock or from iron and steel bar stock obtained from the coastwise trade.

By the 19th century, changing technologies and new strategies for working iron and steel were beginning to impact the nearly ancient iron industries of southeastern Massachusetts. Furnace Village in Easton, MA was among the first (circa 1850) locations in New England where malleable iron was produced. Its owner was Daniel Belcher, the brother-in-law of Seth Boyden, an innovative entrepreneur who reinvented one variation of manufacturing malleable cast iron (Galer 2002). The resulting rapid expansion in varieties of durable, machinable cast iron hardware, combined with improvements in galvanizing technologies and heat treatment techniques such as annealing, gave shipsmiths access to a wide variety of fittings such as belaying pins, braces, clinch nails, and spikes, which could be mass produced in a factory context. Much of the hardware used in shipbuilding soon became the product of the foundry rather than the forge. The period between 1856 and the 1880s was a time of rapid industrial change, even in shipyards such as some of those at Bath, where shipwrights continued to employ pit saws and broad axes for many keel- and ship-framing functions until the late 19th century, despite the appearance of steam-powered woodworking machines (Baker 1973).

By 1856, the importance of England as a source for high quality cast steel edge tools such as plane blades, drawshaves, and carving tools was beginning to decline. While the Sheffield steel industry was reaching its maximum period of productivity in the mid-19th century, the rapid growth of the American iron industry was undercutting reliance on English imports. The forges and furnaces of Pittsburg were already in operation and more were under construction, including crucible steel furnaces. A robust domestic edge toolmaking community utilizing imported and possibly domestically produced cast steel was already well established. The Underhill clan had been producing edge tools in the Boston area since the late 18th century and was now flourishing in New Hampshire; Thomas Witherby was working in Millbury, MA, having arrived in 1849. He was there for several decades before moving to Winsted, CT, taking over both the Winsted Augur Company and the Winsted Plane Company. The Buck Brothers, perhaps America's most famous 19th century edge toolmakers, emigrated from England to work with D. R. Barton in Rochester, NY, before establishing a factory at Worcester, MA, in 1856, and then at Millbury, MA, in 1864. All supplied large quantities of edge tools to the shipwrights of New York, Philadelphia, Boston, and Maine. These companies were beginning to sell their tools in the shipbuilding communities previously served by local edge toolmakers, including shipsmiths. The great florescence of American toolmaking, soon to be exemplified by the classic period of American machinists' tools, was already underway.

Story (1995) has chronicled the history of Essex, MA, one of New England's largest and most unique shipbuilding centers. Though a very small village, Essex built thousands of Chebacco boats and fishing schooners for the Gloucester fishing fleet. He notes that the first steam-powered band saw wasn't adopted in that community until 1884.

> …when Moses Adams set up his steam-powered band saw – the first Essex shipbuilder to do so. At a stroke, so to speak, he revolutionized the building process here by dramatically reducing the labor and therefore the time to get out timbers for a vessel. No longer need frames be beaten out with the broadaxe. (Story 1995, 113)

The coming of the railroad in England, continental Europe, and America and all the industrial development associated with its spread, created a huge demand for iron and steel. In 1840, knowledge of the chemical differences between iron and steel, which was dependent upon its carbon content, was not widespread among smelters and furnace operators. But the technology for producing large quantities of high quality cast iron and wrought iron was already present in the form of blast furnaces, puddling furnaces, and rolling mills. The need for steel rails soon became obvious, and the bulk steel production processes that were invented in the late 1850s became operational by 1870. Mass production of low carbon steel and its use in

Shipsmiths.

Farnham S. & Son,	Bucksport
Sherman John H. & Co.,	"
Crompton Isaiah,	Calais
Vloom Wm.,	"
Welch P.,	"
Devereaux S. K.,	Castine
Hatch & Mead,	"
Godfrey Otis S.,	Cherryfield
Leighton Palmer,	"
Ricker Benj. G.,	"
Cushing A. P.,	East Machias
Chandler Charles H.,	" "
Dennison J. M.,	" "
Gardner C. S.,	" "
Sevey & Chaloner,	" "
Sevey Charles H.,	" "
Buck Amos,	Eastport
Burgin Lewis,	"
Capen David,	"
Whelpley Henry,	"
Bridges J.,	Ellsworth
Stevens & Singer,	"
Harmon Horace,	Lubec
Reynolds Oliver M.,	"
Whalen James,	"
Gross A. K. P.,	Orland
Howes J. R.,	"
McKenney James,	"
Sinclair Benjamin,	Pembroke
Cord Francis,	Robbinston
Gerry Wm. P.,	"
Lamb Andrew,	"
Laughlin Robert,	"
Quinn John,	"
Cooper Andrew,	Sedgewick
Penny Jonathan,	"
Snow Francis A.,	"
Rich Zebediah T., (Bass Harbor) Tremont	
Hatch J. S.,	Waldoboro'
Johnston Francis,	"
Levanseller Ludlow,	"
Mathews Walter,	"
Sides Andrew,	"
Woltz William G.,	"

Figure 51 Maine Shipsmiths from *The Maine Register for 1857 with Business Directory for the Year 1856* (Parks 1857, 279).

The *Weekly Mirror* for 3 March 1854 listed the tonnage built in the several Maine districts in 1853:

District	Tonnage
Passamaquoddy	15,094 – 60/95
Machias	5,495 – 71
Frenchman's Bay	6,850 – 62
Penobscot (Castine)	5,489 – 83
Belfast	9,661 – 43
Bangor	8,531 – 89
Waldoboro	40,453 – 24
Wiscasset	5,011 – 57
Bath	49,399 – 70
Portland	17,549 – 93
Saco	1,635 – 37
Kennebunk	6,590 – 71
York	253 – —
Total	168,918 9/95

[Actual total of above = 172,018 –]

Figure 50 Maine ship tonnages. Table 1 reprinted from William Baker, 1973, *A Maritime History of Bath, Maine and the Kennebec Region*, 2 vols, Bath, ME: Maritime Research Society of Bath, pg. 467.

shipbuilding machinery and for railroads, and the production of hand tools by drop forging techniques are the two most important historical developments impacting, then undermining, the world of the shipsmith, shipwright, edge toolmaker, and the wooden age that had been their milieu for centuries.

New England shipwrights and shipsmiths were particularly conservative in their tenacious clinging to old tools and old techniques. Yet the exquisitely streamlined downeasters of Belfast and Thomaston, the gorgeous three-masted schooners of Waldoboro and Damariscotta, or those sleek four- and five-masted cotton and coal factories under sail built in Bath in the last years of the 19th century wouldn't have been possible without the substitution of steam-powered machinery for hand tools for many shipbuilding functions. The coming of steam-powered machinery wasn't the only factor affecting the shipwrights and shipsmiths of New England.

In 1849, the tentacles of the railroad reached along the Kennebec River to Hallowell just below the head of the tide at the Augusta falls and forever put an end to its heyday as the most important port of central inland Maine. In 1847, the railroad had already been extended to Gloucester, opening up "what amounted to an almost unlimited market for Gloucester's fish" (Story 1955, 60). It took another 25 years for a rail line to be built from Hamilton to Essex, but, when that occurred it was no longer necessary to bring timber from Maine or South Carolina on coasters up the river to Essex. For decades in the early 19th century, shipbuilding timber from New Hampshire had been rafted through a specially built canal in the Plum Island marshes to supply the shipwrights of Essex, who, by the early 1800s, worked in a shipbuilding community where there wasn't a tree in sight. In post Civil War America, the railroad began supplementing the coasting schooner as a principal means of transporting timber to shipbuilding communities such as Boston and Essex. The coming of the railroad had less effect in Maine, where the coasting schooner was the principal means of supplying timber to the Bath, Waldoboro, and downeast custom districts. The vitally important white oak knees had long been cut from the inland woods of Boxford, Massachusetts, near Essex, or the river valleys of southern Maine. After the 1830s many of New England's shipyards would have gone out of business without access to the white oak forests of South Carolina and Georgia, and to a lesser extent to the live oak forests, which extended to Texas and were used in the construction of the highest quality whaling and merchant ships, especially those built between 1820 and 1840 (Little 1971). White oak couldn't be replaced in these shipyards, but traditional tools, such as the broad ax and the pit saw, were being replaced by mechanical equipment by 1882. Not all hand tools were made obsolete by machinery. The smoothing function of the shipwright's lipped adz and slick and the irons and mallets of the ship's caulker could not be replicated by a machine age that soon made shipwrights and shipsmiths obsolete. This was also true of the backing plane, so essential to fit plank and rib, or the finish tools of the ship's joiner for the final fussy work that made shipbuilding an art form, such as small socket chisels, edge-trimming and smooth planes, spar and joiner planes, bevels, calipers, try squares, and hand clamps. These tools are not, nor will they ever be, obsolete, as long as somebody cares to construct wooden boats.

In 1856, however, the railroad had not yet come to eastern Maine. No white oak forests ever grew east of the Kennebec River. A vast undocumented shipbuilding industry had been active in eastern Maine since the beginning of the great migration, after the Treaty of Paris in 1763. By 1856, the flourishing shipbuilding communities to the east of Penobscot Bay, now virtually forgotten, were the location of most of Maine's registered shipsmiths. The tonnage figures reprinted by Baker (1973) from the *Weekly Mirror* (Fig. 51) illustrate the extent of shipbuilding in eastern Maine, the total tonnage of which was nearly equal to that of the city of Philadelphia. Bath and Waldoboro area shipsmiths were now being called by another name, blacksmiths, as exemplified by their presence in the index of Baker's (1973) *A Maritime History of Bath, Maine*. They functioned as shipsmiths, nonetheless, and suddenly reappear in the Maine Business Directory for 1882 (Fig. 59). But there was no shortage of shipsmiths in eastern Maine; the presence of these obscure downeast ironmongers and edge toolmakers is one more lost chapter in the story of New England's tenacious shipsmiths, edge toolmakers, and shipwrights. A most interesting introduction to the milieu of downeast shipbuilding after the Treaty of Paris (1763) opened up eastern Maine to settlement was the author's personal encounter with the oral history of Beals Island and the Great Wass Island archipelago. The story of Manwarren Beal links the late political history of the French and Indian wars, the last historical era before the coming of the steam engine to New England's water mill towns, with the late 18th century settlement of eastern Maine by fishermen, timber harvesters, and shipbuilders, not to mention transient shipsmiths, ships caulkers, and privateers.

Tall Tales from Downeast

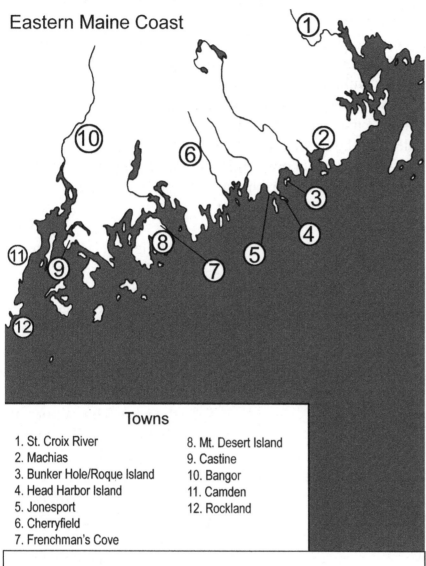

Figure 52 Map showing the Downeast Maine coastline.

Eastern Maine Coast

Towns

1. St. Croix River
2. Machias
3. Bunker Hole/Roque Island
4. Head Harbor Island
5. Jonesport
6. Cherryfield
7. Frenchman's Cove
8. Mt. Desert Island
9. Castine
10. Bangor
11. Camden
12. Rockland

According to oral history narrated to the author long ago, before Manwarren Beal, the first settler on Beals Island, ME, came there to live, he first landed on the north side of Head Harbor Island at the village on the cove on Seguin Passage. Though not labeled in the *Maine Atlas and Gazetteer* (2005), this cove is labeled "Beals Cove" in the *1881 Atlas of Washington County*. On Manwarren's first trip to Head Harbor Island, he and his trading (?) privateering (?) crew were first attacked by Indians on the island itself. Later that night while anchored in Beals Cove due to stormy weather, Manwarren thwarted a second Indian attack when, hearing a noise at the stern of his vessel, he "leaned over and hooked the fish gaff into the Indian's jaw, and began pulling him aboard as he would a large fish" (American History Department of Jonesport-Beals High School 1970).

Manwarren was born on September 12, 1736 and died on August 23, 1800. He was presumably visiting Head Harbor Island just before the ending of the French and Indian Wars (1759). He returned again to Head Harbor Island before choosing a location to build his house on Beals Island. The exact date of his arrival is unrecorded, but he was part of the great influx of settlers to the Pleasant River settlements, which began shortly after the Treaty of Paris in 1763. This treaty signaled the end of the French and English conflict over the control over large sections of North America, including the maritime peninsula. How long Head Harbor Island had been settled before Manwarren Beal immigrated from Martha's Vineyard is unknown. Nearby islands, such as Roque Island and the mainland tidewater areas east of Mount Desert Island, had long been the domain of French settlers, since the early 17th century. Frenchman Bay in general, and Frenchman's Cove in particular (now Hulls Cove, once part of Eden) had, due to its protected harbor and year-round fresh water (Breakneck Brook), long been a rendezvous for French fur traders and indigenous First Nation Abenakis crossing the bay from the Jordan and Taunton rivers to trade with the French. Head Harbor Island may also have been a trading station location, but the principal commercial activity taking place on the island at this date was probably the trade in and/or exchange of goods confiscated by the many privateers who operated along the New England coast.

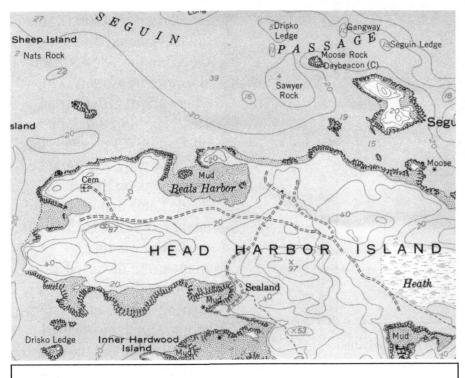

Figure 53 Beal's Harbor. USGS survey map.

At the time Manwarren Beal came to Beals Island, Gloucester fishermen sailing their Essex-built Chebacco and dogbodies were settling Mt. Desert Island and Deer Isle to the west. The capture of Quebec in 1759 and the end of the French and Indian wars opened the entire coast of Maine east of Pemaquid to fishermen and their families. The threat of further harassment from bands of Abenaki from the St. Lawrence River communities, who previously used the Norumbega hill country northwest of Wiscasset and Thomaston as a base for harassing coastal settlements, ended with the Treaty of Paris in 1763.

It is not known when the first shallop, pinky, dogbody, sloop, or schooner was built at Frenchman's Cove, where the Brewer family later built their coasting schooners (±1813). Frenchman's Cove is located 40 miles west of Head Harbor Island but the start of shipbuilding all along the downeast coast from Mount Desert Island to Passamaquoddy Bay was probably contiguous with the great wave of settlers who came to the area after 1785. Further downeast at the bustling fishing communities at Machias, Machiasport, Jonesport, Addison, Columbia Falls, Harrington, Cherryfield, and Milbridge, the first order of business would also have been shipbuilding. Coasting vessels from lumber- and hay-starved Martha's Vineyard, Boston, Salem, Marblehead, and elsewhere had long been coming downeast for firewood, cordwood for clapboards and staves, timber, masts, spars, hay, and salted codfish. By 1770, shipbuilding was already well underway at Warren, Thomaston, Castine, and probably at Sedgwick and nearby tidewater communities. The fishing and coasting vessels of the time were variations of the Chebacco boat, the smaller vessels pinkies and dogbodies; the larger vessels schooners or proto-schooners. Some of the Chebacco style boats brought to Maine by the flood of settlers after 1785 were probably built at Essex, MA, where more fishing boats and small coasters were constructed over an almost 300 year period than in any other location in North America. But such watercraft were also built in downeast Maine, the majority of which were not the subject of any recordkeeping or notations in any local town histories. Someone had to iron these ships, and to provide the edge tools to build them. Scattered stories survive that tell us fragments of the maritime history of downeast Maine, as with the tales of Manwarren Beal. Some tools also survive from the downeast shipbuilding era, which continued for a century after 1785, but downeast shipsmiths are almost entirely forgotten.

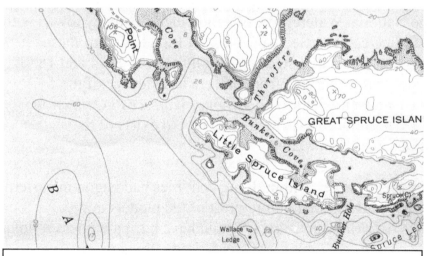

Figure 54 Bunker Cove. USGS survey map.

The early history of Head Harbor Island has been lost; some might contend that it never existed, at least not until the granite industry arrived in the 19th century. But the oral history of Beals Island has it that ships were built at Head Harbor Island as soon as the first privateers arrived from Martha's Vineyard. During the later years of the Indian Wars and the French and English contest for east Maine, not all the isolated and, in many cases, undocumented settlers were French. The inaccessibility of Head Harbor Island afforded protection from the marauding indigenous natives allied with the French who still traversed the area until the late 1750s. When the later flood of Revolutionary War veterans began migrating to the Pleasant River settlements, an isolated community of English descent, privateers and fishermen, may have already been established at Head Harbor Island. If they built or repaired a fast schooner, a takeoff of the already famous Baltimore Clipper so essential for their trade, where did they get their iron fittings and especially their tools? What type of sailing vessel brought Manwarren Beal and his fellow immigrants to east Maine, and what kind of vessels were used by the settlers or privateers who greeted them? When Manwarren Beal came to Beals Island, did he come to fish or to harvest cordwood or timber for the many shipbuilding communities to the west or was he a privateer who decided to settle down on Beals Island? Did he have a vessel that he kept in Bunker Hole, that famous hiding place behind the cliffs at Spruce and Great Spruce islands for the privateers who ranged the Atlantic coast for over 200 years until the end of the War of 1812? Did he build his own coasting vessels or get one by hook or by crook from the shipyards of Essex? How did he "make do?" Where did he get his tools and who made them?

In *The National Watercraft Collection*, the majority of surviving half models in the Smithsonian collection listed as merchant sail (mostly schooners and brigantines) are models for ships built at Newbury (Newburyport) shipyards along the Merrimac River or were constructed in the area lying between Belfast and Passamaquoddy Bay, Maine (Chapelle 1960, 55, 60-88). No Bath and other Kennebec River half models of vessels, built in huge numbers and meticulously documented by Baker (1973) and others, are in this section of the half model collection. The listing of Penobscot Bay and downeast communities represented in the National Watercraft Collection includes many towns with no current reputation as shipbuilding centers (See Appendix VI for a listing of these downeast built ships). The ships that Chapelle listed were built between 1844 and 1880; to be in the collection, half models had to survive from this era as evidence of the construction of these ships. The earliest model in *The National Watercraft Collection* is the *Atticus*, built at Castine in 1818. Before 1840, almost no record exists of the shipbuilding activities at locations in eastern Maine.

As a matter of community survival, however, shipbuilding activities had begun in eastern Maine with the coming of the first settlers, long before most of the models in the Smithsonian collection were laid out. Manwarren Beal would have been a witness if not a participant in a vibrant late 18th century community of shipwrights, shipsmiths, privateers, fishermen, and merchant adventurers. Downeast Maine would have been an important destination for coasting vessels due to its rich natural resources and ease of access via prevailing southwesterly winds. The Ruggles House in Columbia Falls (c. 1818) is a testament to the flourishing mercantile community that probably reached its maximum period of prosperity during the neutral trade and prior to Jefferson's embargo of 1808. The highest population levels of Euro-American settlers in eastern Maine were reached in the first three decades of the 19th century. A vigorous shipbuilding trade was well established by this time. Jefferson's Embargo Act of 1808 and the War of 1812 drastically, but temporarily, curtailed shipbuilding in the region. The half models in *The National Watercraft Collection* from eastern Maine represented not a shipbuilding industry that suddenly arose after the 1830s but one that had its birth before the American Revolution. The surge of shipbuilding in coastal Maine in the 1830s and 1840s is well documented; activities before this date are nearly undocumented. Maine's vigorous shipbuilding industry was rooted in a century and a half of the construction and "ironing" of ships in southern New England, the center of such activity in the colonial era.

Figure 55 Cooper's shave made by G. B. Ricker of Cherryfield, Maine, 15 ¾" long, in the collection of the Davistown Museum 62504T1.

In the 18th and early 19th centuries, extensive timber resources suitable for both southern New England and Maine shipbuilders would have been located just upstream from the tidewater communities of eastern Maine. At this time, southern New England was continuing to suffer from a scarcity of firewood, which began in the early 18th century. The timber for masts and spars and white oak for ship's knees located inland from Essex and along the shores of the North River in Massachusetts had been depleted by the turn of the 19th century. The residents who moved to eastern Maine had more reliable and profitable opportunities to earn their livelihoods than by privateering. Fishing had always been the most important trade for New England's coastal residents, but the need for timber, cordwood, masts, spars, and hackmatack, the most common substitute for southern New England's now depleted white oak timbers, inspired a tenacious post-Revolutionary era coasting trade in Maine, which endured for another hundred years. So too, did Maine's shipbuilding trade, which evolved hand-in-hand with the growing coasting trade of the late colonial era and the early republic.

Surviving tools made by Maine toolmakers and shipsmiths in the early and mid-19th century, in some cases one and the same person, provide vivid testament to the extent to which Maine communities relied on locally made edge tools. Unlike edge tools, Maine shipsmiths did not sign their names on the forged ironware they made to iron the ships they built, but they did sign their tools and often noted the location of their forges. These tools survive today as primary evidence of a vigorous edge toolmaking community working in or near all the shipbuilding centers of coastal Maine. In many cases, their signed tools are the sole evidence of the activity of edge toolmakers who were not always listed in Maine business directories or town histories. The same observation can be made about southern New England toolmakers and those working in other states at this time.

There are bog iron deposits in coastal Maine, especially downeast in the highlands above Cherryfield, Harrington, and Machias. No records exist, however, of any bog iron-smelting activities in these areas. The Rickers of Cherryfield were well known edge toolmakers who also forged iron fittings for locally built ships. Many of their edge tools, which vary from raw, slag-laced, heterogeneous natural steel to the sleek homogenized shear steel as in the drawknife illustrated (Fig. 55) survive to remind us of the longevity of downeast shipyards and the probable use by local shipsmiths of nearby bog iron deposits. The extent to which they used bar steel brought by coasting vessels from steel furnaces in southern New England or Pennsylvania to "steel" their edge tools, or forged their own steel by steelmaking strategies long remembered from ancient times is now one of the mysteries of an earlier age. The wide variations in the appearance of Ricker tools, in particular, attest to a range of production strategies and diverse sources for the iron and steel used in their tools.

In many cases, as with the Rickers of Cherryfield in Downeast Maine, the forging of edge tools was also accompanied by the production of ship fittings. Benjamin Ricker is listed in the 1856 Maine Business Directory as a shipsmith; also listed as shipsmiths working in Cherryfield were Otis Godfrey and Palmer Leighton. Numerous specimens of Ricker signed edge tools are in the collection of the Davistown Museum and attest to the productivity of the Ricker clan as well as that of the Cherryfield-Harrington area shipyards of the early and mid-19th century. The fact that Amaziah Ricker's federal style house, still present in Cherryfield, was built in 1803 provides evidence that shipbuilding, edge toolmaking, and shipsmithing began with the great migration to the Pleasant River settlements in the years before the Revolutionary War. The well documented history of the Ricker family (Cherryfield town library records) is the exception to our observation that almost all information about late 18th century and early 19th century shipsmiths and shipbuilding in downeast Maine between the great migration of the 1760s and the boomtown years of lumbering in Bangor, Ellsworth, and elsewhere in Maine, 1820 to 1840, has been lost.

Lost and Found History

History is the narration of the events and accounts of the past, often but not always in the order of time. As noted by Schlesinger (2005), the job of the historian, is, in fact, to rewrite history, altering old narrations.

Implicit in our endeavor to rewrite some chapters in the industrial history of colonial New England is the concept of history as palimpsest, i.e. a parchment containing a narrative which has been erased or covered over by subsequent narrations. Part of the rewriting of history is also the uncovering and disclosure of facts and stories not previously recorded. Modern historians usually mark their success by their recounting of facts and accounts not previously known, including their exploration of new primary sources, which alter or discount and, in some cases, erase earlier narrations. But history is not only a process of reinterpretation of old narratives, soon to be erased by the inscription of new texts. It also includes the constant dynamic of forgetting or laying aside old accounts and narratives. Many narrations are not reinterpreted but simply lost, forgotten, or abandoned.

The recounting of historical events must include consideration of the significance of material cultural artifacts in museums and historical society collections as essential components of any historical narrative. Among these surviving material cultural artifacts are not only the documents, diaries, and written histories, but the tools (accidental durable remnants) used for the essential commercial activities of these communities. In the case of accounts of societies and cultures in prehistory, surviving material cultural artifacts, often in the form of accidental durable remnants found in archaeological excavations, are the primary basis for the narration. The rediscovery of the accidental durable remnants of the past, the very essence of the interface of lost and found history, provide a unique opportunity not afforded by the written word to enlarge and reinterpret our understanding of the past. No accidental durable remnants of past history tell us more about colonial history than the hand tools that built the infrastructure of the wooden age. The ships built by shipwrights with the help of shipsmiths and edge toolmakers have long since rotted away. Only written, painted, or etched accounts tell of the existence of these ships prior to the introduction of the camera. But written accounts are selective, if not myth-making, leaving out essential information about the milieu of an era, its steel- and toolmaking strategies and techniques, and the individual artisans who were the creative enablers of the era we seek to reinterpret. The tools themselves are important additional primary sources of information about our colonial maritime history, voices from the past narrating stories not contained in contemporary written texts.

In this context, our sketch of the New England shipsmiths, the origins of the steel and iron used for their edge tools, and the legacy of their contribution to American culture is a constant confrontation with the lost, forgotten, or undocumented chapters in our industrial and maritime history. We can explore, at least in a cursory sense, known chapters in the local history of the North River, the Taunton River watershed, the Carver blast furnaces, or the shipwrights of Old Rochester. We can go back in time to the beginning of our narration symbolized by the construction of the *Virginia* at Fort St. George (1607) and the Saugus Ironworks (1646), and explore the nature of ferrous metallurgy and edge toolmaking before and after this period. But a vast territory of unexplored or forgotten history lies between early colonial iron-making and the rise of the American factory system in the mid- and late 19th centuries. The huge watershed of the Merrimac River and its industrial history is one example of regional toolmaking and shipbuilding worthy of a voluminous history, as is the history of shipsmithing and shipbuilding on the Piscataquis River. Significant bog-iron-smelting must have occurred in Connecticut as well as New York, New Jersey, and the Chesapeake shore as a component of colonial era shipbuilding in these areas. The huge shipbuilding output of Mystic, Connecticut, may have close links with the Taunton River smelters, especially in the colonial era, but that chapter on shipsmithing has yet to be written. Beachcombers on Martha's Vineyard and possibly on Nantucket Island collected bog iron from the shoreline below eroding cliffs, as well as from bog iron swamps on Martha's Vineyard, and shipped it by coaster up the Taunton River (Flenders 2007, personal communications). Their interest in Taunton as a location of forges whose owners reimbursed them for the trouble of collecting bog iron and transporting it up the Taunton River reminds us of the total reliance of both Nantucket whalers and Martha's Vineyard merchantmen (and privateers) on the mainland, principally Buzzards Bay, but also North River valley, shipyards, shipsmiths, and inland as well as imported timber resources, ironwares, and tools.

That residents of Martha's Vineyard and possibly Nantucket walked the beaches of the south shorelines of these islands and collected bog iron for the bloomsmiths of the Taunton smelting furnaces is an oral historical account on the verge of being lost. More well documented are the narrations of the origins and growth of shipbuilding in Buzzards Bay, the location of construction of many a Nantucket and New Bedford whaler (LeBaron 1907). Even here, the story of the earliest bloomsmiths and toolmakers lingers on the edge of lost and found history.

In the case of Mattapoisett (Old Rochester,) Buzzards Bays' most productive shipbuilding community, several local histories make brief reference to "Leonard's Sippican forge," which was established after the end of King Philip's War, the termination of which opened up the shores of Buzzards Bay for settlement and shipbuilding after 1680. Along with its more well known neighbor, the first corn mill, that other first necessity of early settlers, the Leonard forge on the Sippican River may

148

have been in operation by the late 1680s, and certainly by 1703, probably at the head of tide, always a popular choice for siting water-powered forges and mills (Leonard 1907b).

Figure 56 Mortising chisel, natural steel, signed "KIMPTON" with a backwards N and a scalloped edge around the imprint, also with a first initial that might be "I" or "J". In the Davistown Museum collection 080907T1.

Even before Leonard's Sippican forge was smelting bog iron blooms, Ezra Perry was forging ship's hardware and edge tools on the opposite side of Buzzards Bay adjacent to the Aptucxet Trading Post, which the Pilgrims had established by the mid-1620s. Some of the tools produced by late 17th and early 18th century shipsmiths and edge toolmakers working at this time (unlike the wooden ships these tools built) still survive from this era. Lost, then found, tools surviving from the early colonial period tell stories about the florescence of shipbuilding in colonial New England, which are only briefly referenced in town and regional histories, if at all. Dozens of signed American-made edge tools have been recovered by the Liberty Tool Co. from New England cellars, workshops, and tool chests and now reside at The Davistown Museum. One example of an early 18th century mortising chisel (Fig. 56) has been recovered from a tool collection associated with New Bedford-Old Rochester shipbuilding in the mid-19th century. The context of the use of this tool as part of a shipwright's tool kit is clearly evident, but this tool predates the heyday of whale shipbuilding at Old Rochester by a century. This mortising chisel is stamped with the mark "J. KIMPTON". Who was this toolmaker and where did he forge this tool? This question can be repeated ad-infinitum with respect to edge tools continuously being discovered that are not listed in the 1999 edition of the *Directory of American Toolmakers* (Nelson 1999). The Kimpton mortising chisel is only one example of the reoccurring interface of lost and found history. Perhaps further research will show that Kimpton is one of the many undocumented edge toolmakers of the Taunton River watershed working in the late 17th or early 18th centuries making tools used by shipwrights such as Thomas Coram (Dighton, MA, 1697 – 1702). This mortising chisel typifies domestically produced edge tools made by shipsmiths during this period of booming southern New England shipbuilding activity.

Another example of the dynamic interface between lost and found history are four edge tools marked H Merritt, a lathing hatchet, broad ax, hewing ax, and timber-framing chisels now in the collection of the Scituate historical society and on display at the Cudworth House Museum. Who was H. Merritt was and where did he produce the tools used in Scituate shipyards? The Merritt tool cache is 19th century in style and metallurgy; he may have been a North River shipsmith or one of the many edge toolmakers working in the nearby Blackstone River or Merrimac River watersheds. There is no record of any H. Merritt in the principal reference on American toolmakers working before 1900, (Nelson 1999). Nor is he mentioned as a toolmaker in Briggs' (1889) survey of shipbuilding on the North River. Surprisingly, Briggs also fails to mention the famous and now well known spar planemaker, J. R. Tolman, who is only mentioned in Briggs as the seller of part interest in "Mr. Fosters warship to Samuel Tolman Jr." on January 11, 1817 (Briggs 1889, 233), but Briggs does list a family of shipbuilders beginning with Captain Benjamin Merritt who settled in Scituate in 1709. Briggs lists Henry Merritt as a shipwright who is noted as building the *Sarah Jane* (1851) in the twilight years of North River production (Briggs 1889, 233). It is possible that the principal occupation of H Merritt may not have been shipbuilding but rather shipsmithing and edge toolmaking and that the tools now on display at the Cudworth House Museum could be the only known surviving tools of an obviously accomplished toolmaker.

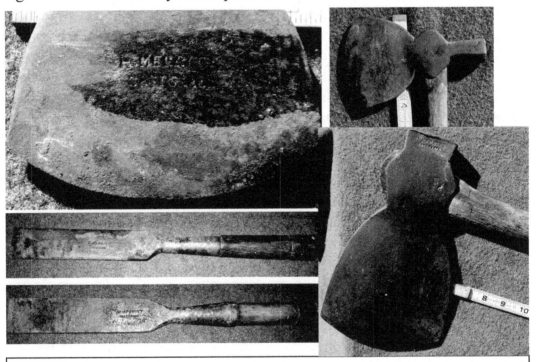

Figure 57 H Merritt tools at the Scituate Historical Society. Photographs compliments of E. Decker Adams.

The Scituate Historical Society also has a hewing ax made by James Howland of Bucksport, Maine, noted as working before and after 1879 (Brack 2006). But this ax is also signed Stephanson, suggesting that another smith at some unknown location, but probably not Bucksport, later repaired this tool by reforging it, a common practice for saving the cost of forging or buying a new tool (Smith 2007, personal communication). It was the job of the multi-tasking shipsmith to repair by reforging the worn out tools of the shipwright. In early 19th century Mattapoisett (Old Rochester), numerous blacksmiths

150

(Mendell 2007) are noted as working up the side streets from the major shipbuilding yards along Water St. Using forging techniques unchanged for centuries, these shipsmiths made their daily visits to shipyards to "iron" the whaling ships and merchant fleet first built in the 18th century for Nantucket whalers and Martha's Vineyards traders and later for whale ships of Old Dartmouth and New Bedford after the harbor at Nantucket silted in. An excellent example of a typical blacksmith's forge is on display at the Mattapoisett Historical Museum. This bellows-driven forge is representative of those used by shipsmiths to forge the muck bars of the local bloomsmith into ship's hardware, as well as for forging the iron bodies of edge, horticultural, and other hand tools used in the local community.

Paul Rivard is of particular interest to students of Maine history who are interested in the interface and dynamics of lost and found history. Rivard is a noted historian associated with the Maine Historic Preservation Commission. In his review of Maine's industrial history (Rivard 2007), Rivard fails to mention not only the ironware produced by Maine's numerous shipsmiths but also the many shipyards that produced Maine's sailing ships in the 19th century. As an example of our tendency to forget some of the most interesting chapters in local and regional history, Rivard's text and index also contain no reference to shipbuilding, shipsmiths, or shipbuilding tools. The output of individual shipsmiths, including Maine's edge toolmakers, may now seem insignificant, but their productivity in

Figure 58 Bog iron claw hammer, 6" long, 1" square face, 10" wooden handle in the collection of the Davistown Museum TAB1003.

the early 19th century represents a link between the earlier accomplishments of southern New England ironmongers and edge toolmakers, the success of the American Revolution, and the later evolution of a vigorous American maritime and industrial economy. As edge toolmakers working in America's most productive shipbuilding state (after 1840), Maine toolmakers have been overlooked by historians, but they were an important component of the evolution of the American factory system. The hand tools they forged were not replaced by factory-made tools until the last two decades of the 19th century.

Maine's relatively late preeminence in shipbuilding and its more obscure accomplishments in the art of edge toolmaking are only a fragment of the vigorous productivity of the classic period of New England's industrial renaissance (1830 – 1930), which is rooted in the colonial legacy of ironworking and toolmaking. These colonial traditions are, in turn, grounded in centuries of iron mongering in continental Europe and England, which preceded the construction of the pinnace *Virginia* at Fort St. George. How many thousands of local New England farm family forges and smithies used smelted bog iron bar stock from direct process bloomeries to forge tools for their own use or for that of the communities in which they lived? How often were farmyard forges both the bloomery for smelting the bog iron and the forge for making tools, in the long tradition of iron-smelting that dates to the earliest days of the Iron Age? The primitive forge welded claw hammer found in Plymouth county (MA) in the early 1990s (Fig. 58) is an excellent example of an early colonial era hand tool made from unrefined bog iron. How many village blacksmiths, whom we now think of as farriers shoeing horses, were adept at the multitasking colonial necessity of smelting bog iron, then making horticultural tools, ships' fittings, and edge tools for farmers, shipwrights, barn builders, house wrights, and other timber-framing uses, and then finally shoeing the oxen who were the most important colonial transport medium? What record do we now have of their role in the evolution of a colonial maritime economy into the vigorous American industrial economy of the post-Civil War era, other than the tools and iron fittings they left behind in old tool chests and dark cellar workshops? Written records and accounts detailing the activities of these and other ironmongers are either lost or nonexistent. Many of the tools that survive from this era as accidental durable remnants (ADR) allow a better understanding of and new narrations about colonial and early American history.

As interest in colonial toolmaking and shipbuilding declines, so too does the search to uncover the details of our historical past. Once relatively well known facts about the identity of essential community artisans and the products they produced or sold, if not recorded in old town histories and other records, are lost with the death of the witnesses to the prosperity and then decline of these communities. Lawrence Norton, the town historian of West Jonesport, Maine, never wrote down his knowledge of the commercial history of West Jonesport (now gone – deconstructed almost in its entirety) that he retold to this author. No town history exists providing the detailed recollections he recounted, yet the Jonesport town history (American History Department of Jonesport-Beals High School 1970b) contains one circa 1850 photograph of a four story hotel, the Bayview Hotel, which burned in 1872, located behind Lawrence's house. Even Lawrence Norton couldn't explain why such an imposing structure was built for visitors in the early 19th century in West Jonesport. This author suspects that it was for visiting shipwrights and ships' caulkers working in the undocumented shipyards of West Jonesport and the Pleasant River at Addison, who were known to sail from community to community to build one ship and then another in a nearby port.

Our search for information about the tools recovered by the Liberty Tool Co. brings us face to face with the mysteries and dynamic reality of the phenomenology of history. Our first observation is that history is not a straightforward series of events that can be easily documented and narrated. In reality, disorderly history is an unruly, overlapping mixture of remembered and recorded history often masquerading as scientific fact, written documents and memoirs, oral history, including stories and myths, and material cultural

152

artifacts, including tools. Surrounding the world of factual history like the rings around the planet Saturn are more ephemeral components of the phenomenology of history: protohistory (the written observations by literate observers of communities who did not themselves record their history), prehistory (documented primarily by the accidental durable remnants of archaeology) and lost and forgotten history, often but not always beyond the realm of rediscovery.

In this context there is a massive body of written history that narrates stories or records facts pertinent to the tools we discover and seek to understand. But found tools as accidental remnants also confront us with stories, events, and steel- and toolmaking strategies and techniques that we don't understand. Every hand tool has its own history – of ownership, use, historical context, fabrication, and loss. The compilation of the *Hand Tools in History* series is our attempt to understand what our hand tools are telling us, i.e. the phenomenology of tools: the stories of not only how and why we use and have used tools but the discovery of what tools tell us about the past as well as the future.

The tools themselves are the primary evidence of their forging, of their intended use, of the communities that depended on their function, and of the individuals who forged them, used them, or cast them aside. The metallurgy of the tools illustrated in this volume discloses questions that are not readily answered by written historical texts or technical reference material. No more intriguing historical questions exist for us than those elicited by the tools used for the construction of the pinnace *Virginia* and other 17th century ships. This era, the early colonial period of New England's maritime history, is a link between the fascinating steel- and toolmaking strategies and techniques of prehistory, the early Iron Age, the migration and medieval periods, the high Renaissance of European world exploration and conquest, and the modern period of our consumer product-centered growth oriented pyrotechnological existence.

In that unscientific, ephemeral milieu of lost and found history, oral stories and anecdotes are passed from one generation to another, and incomplete written histories hint at lost chapters not recorded. Accidental durable remnants in the form of found tools and ironware join with anecdote and oral history to remind us of the dynamic interaction between the forgotten and the remembered, an essential element in our attempt to understand the past. Implicit in our attempt to peel back the palimpsest of history to disclose the events, skills, and traditions of the past, is that any attempt to retrieve lost history inevitably involves our subjective interpretation of the past. Even steeled edge tools, which theoretically can be subject to incontrovertible metallurgical analysis, can be the subject of myth-making. The tools themselves are, however, a primary source of information about the historical past and beckon us to reconsider and reinterpret forgotten chapters in colonial and early American history. The uncovering of lost history: one of the pleasures of the text.

Shipsmiths of 1882 – the End of an Era

Ship Smiths.

(See also Blacksmiths.)

Cassidy T. F., Front, opp. May, Bangor
Kelley E. J. 12 Washington, "
Corliss William E., Broad st. Bath
Crooker & Lilly, Commercial st. "
McQuarris John, Water st. "
Potter Elisha, Water st. "
Wetherbee Timothy, Water st. "
Kavanagh Frank, Main st. Brewer
Russell Dexter W. Camden
SIDELINGER G. B., Rockport, "
Hatch Augustus, Damariscotta
AVERY & McLAURIN, 17 Union (see advt. p. 415), Portland

Laughlin Thomas & Son, Commercial, cor. Centre, Portland
Mayo Albion W., Railroad st. "
Morgan Patrick, 285 Commercial, "
Stanwood Charles, 262 Commercial, "
STANWOOD GEORGE M. & CO. 261, 263, and 265 Commercial (see advt. p. 12), "
Taylor William, 54 Commercial, "
Atkins W. J. 16 North Main, Rockland
Hahn S. B. Thomaston
Hanley George, Water st. "
Matthews O. D. "

> Figure 59 Listing of shipsmiths from the Office of the Librarian of Congress, 1882, *The Maine Business Directory, for 1882.* Boston, MA: Briggs & Co., Publishers, pg. 282.

The Ricker family of shipsmiths, blacksmiths, and edge toolmakers has lived in Cherryfield, Maine, for over 200 years. The survival of a Ricker family blacksmith/shipsmith shop into the 21st century is a reminder of how the historical documentation of many other seemingly unimportant toolmakers and shipsmiths has been lost to us. The Ricker family forges and workshops probably included more shipsmiths and edge toolmakers than just George Benjamin Ricker. The longevity of their operations in Cherryfield and the large quantities of tools produced over decades of smithing attest to their important role in downeast shipbuilding.

The Ricker clan is thus a remembered component of our admittedly erratic sketch of shipbuilding in Downeast Maine. As both the listing of the 1882 shipsmith's working in Maine and the many wooden ships still being built at this late date attest, an active shipbuilding industry was still flourishing in coastal Maine as well as at Essex, MA, and other southern New England locations. But rapid industrial change was underway, best symbolized by the huge Bessemer pneumatic and Siemens open hearth steel furnaces now at work smelting steel in multi-ton batches. Henry Clifton Sorby was just about to publish his two decade old microscopic images of the crystal structure of wrought iron, steel, and cast iron. The science of metallography was beginning to catch up to the galloping technology of making iron and steel in an industrial age where Maine's shipsmiths and edge toolmakers were already a forgotten footnote to New England's flourishing mixture of textile mills, shoe factories, tool manufacturers, and ever present maritime fisheries and forestry industries. The 1882 listing of Maine shipsmiths is the last listing of this obscure trade in Maine business directories, providing a convenient year to end our exploration of only a few of the pathways in the labyrinth of the history and the art of the edge tool.

Encapsulated between the success of the American Revolution and the publication of the last 1882 listing of shipsmiths in Maine is the rise of a vigorous indigenous American iron-, steel-, and toolmaking industry and a century of wooden shipbuilding partially documented by New England's many maritime museums. Individual town histories and maritime historians ranging from famed writers, such as Howard Chapelle to more obscure town historians such as Cyrus Eaton, document ships designs and construction, cargoes, trading routes, destinations, the social milieu of maritime communities, and tales of captains, crews, and shipwrecks. Less well documented are the tools that built these wooden ships; even more obscure are the steelmaking strategies and toolmaking techniques that gave birth to the tools themselves. Entirely forgotten are the shipsmiths and edge toolmakers who made the tools that built the ships that ensured the success of the American Revolution and the growth of the vigorous mercantile, maritime, and industrial society which followed. Most compelling of all is the stark presence of these tools, often found in the disorder of old wooden tool chests or the dark cellars of New England Victorian houses. A south Bristol, Maine, boat shop, a Quincy Massachusetts, workshop, frost fish spears under a dory at Acushnet, all locations of common hand tools that upon close inspection disclose stories about the roots and florescence of American toolmakers and the role they played in the maritime and industrial history of New England. The oldest of these often forge welded artifacts, ironware as well as hand tools, remind us that, for almost 150 years preceding the Revolutionary War, a vigorous indigenous community of artisans living along or near the New England coastline was smelting iron and making tools, laying the groundwork for the coming revolution.

The ability of early colonial shipsmiths to "iron" ships with bog iron derived fittings was soon followed by the evolution of a robust indigenous edge toolmaking community. These industries were essential to the later success of the American Revolution and constitute an essential prologue to the later growth of New England's shipbuilding industry in the 19th century and the concurrent development of the American factory system of tool and gun production. The finesse of New England ironmongers, and in the 19th century Maine edge toolmakers, is an obscure but important link between early tool- and steelmaking techniques and strategies and the massive production of America's classic period of hand tool production (1870 - 1930).

Our commentary on New England's iron smelting, shipbuilding, and toolmaking industries ends with the advent of modern science, the chemistry of lattice structure, iron carbon diagrams, eutectic points, and the evolution of modern steelmaking technologies. The creative output of the shipsmiths and edge toolmakers who made the tools that helped build America is based on the combination of the science and art of metallurgy, i.e. the historic marriage of C. P. Snow's (1959) two cultures, which ended with the demise of the wooden age of the shipwright and the onset of the industrial society now in its twilight.

Now forgotten New England shipsmiths and edge toolmakers were important participants in the rapid growth of New England's flourishing maritime economy from the end of the great migration to Massachusetts Bay (1643) to the heyday of shipbuilding in Massachusetts and Maine in the mid-19th century. The legacy of their metallurgical finesse can still be seen in the form of the forge welded shipbuilding tools – now the primary evidence for their existence – that still survive despite the fact that the ships they built were long ago lost at sea or to the biological hazards of rot and sea worms.

These tools keep disclosing themselves as accidental durable remnants in a glowing landscape of industrial debris, paved wastelands, and chemical fallout. The hand tools themselves are a reminder of the evolution and growth of a resource devouring not-so-convivial pyrotechnical society. The irony of the florescence of American toolmaking and its classic period of American machinist and woodworking tools lies in the conjunction of objects of great beauty, artistic craftsmanship, and creative usefulness in the context of the flowering of an industrial society of fleeting presence, which quickly devolved into a post-industrial consumer society characterized by toxic plastic and electronic products and rapidly accelerating ecosystem collapse and the prospect of cataclysmic climate change. The temporary ascendancy of this polymetalic pyrotechnical industrial society was in part due to the success and finesse of bloomsmiths, shipsmiths, edge toolmakers, and shipwrights who were the incarnation of the marriage of art, tools, and history. The irony of their construction of vast fleets of schooners, a distinctly American art form, is that their evolution into bulk cargo carriers of ever increasing size helped put an end to the wooden era that was engendered by their skills and their sculptures under sail. The construction of the pinnace *Virginia* at Fort St. George in 1607 is the first chapter in the complex history that constitutes our maritime and industrial history.

That last century of the finest wooden ships ever built rested on three centuries of tenacious indigenous iron-smelting, edge toolmaking, shipsmithing, and shipbuilding by the residents who lived along the New England coast from the shores of Connecticut and the bog iron country of southeastern Massachusetts to downeast Maine and the land of Manwarren Beal, Bunker Hole, and the far eastern shipbuilding centers of the Passamaquoddy customs district, now forgotten except for obscure listings in old business directories and half model collections. These forge masters, founders, blacksmiths, shipsmiths, edge toolmakers, and shipbuilders were the forefathers of the florescence of American toolmaking that occurred in New England in the 19th century and spread west with the railroads and America's growing population. New England is but one region in a nation with many other toolmakers and shipbuilding communities whose roots lie in centuries and even millennia of shipbuilding, trading, world conquest, and edge toolmaking. The only evidence of the robust communities of artisans who made the tools that built the wooden ships that long ago disappeared are the accidental durable remnants of this earlier period in American history, now found in dark corners of old workshops and unused tool chests. These are the tools that, in fact, built a nation.

Appendix A

Definitions: Types of Iron and Steel

The following definitions are used throughout the *Hand Tools in History* publication series. See *Handbook for Ironmongers: A Glossary of Ferrous Metallurgy Terms* for a more detailed description of the diversity of definitions of iron and steel.

Wrought iron: 0.0 – 0.08% carbon content plus slag inclusions

Malleable iron (modern definition): 0.08 – 0.2% carbon content plus slag inclusions

Malleable iron (pre-1870 definition): 0.08 – 0.5 carbon content

Low carbon steel (post-1870 definition): 0.2 – 0.5% carbon content, no slag inclusions

Tool steel: 0.5 – 1.3% (– 2.0%) carbon content

Cast iron: 2.0 – 4.5% carbon content

There is a wide variation in antiquarian and modern definitions of iron and steel. Before the era of modern bulk process steel, "malleable iron" included what would now be called low carbon steel and can also mean any iron with a carbon content up to 0.5%, which is the hardening and quenching threshold for steel. Malleable iron contained varying amounts of siliceous slag; more slag was present in direct process bloomery derived iron, less in indirect process (blast furnace) derived puddled iron. In the modern era, the definition of low carbon steel is so broad as to range from 0.08 to 0.5% carbon content.

There is also a gray area between steel with a carbon content above 1.5% and cast iron at 2.2%; few, if any, tools have higher carbon content than 1.5%, and most cast iron and malleable cast iron products have a carbon content above 2.2%. The *Encyclopedia Britannica* defines steel as having less than 2.2% carbon content; other sources, such as Gordon (1996), define the limit at 2% carbon content. This text follows the latter definition.

Modern writers often call all iron that is not steel "wrought iron." This is misleading; there is a great difference in the hardness and durability of tools made of malleable iron with a 0.3 or 0.4% carbon content and tools made of wrought iron with 1/10th as much carbon content. The latter artifacts would be much more susceptible to plastic deformation; in both cases, slag inclusions increased tensile strength and added qualities of ductility completely lacking in slag-free modern low carbon steel. Both wrought and malleable iron ceased to be produced in any significant quantities after 1930 as a result of the demise of the puddling furnace. Low carbon "steel" became the dominant form of iron produced after the advent of the Bessemer and open hearth bulk steel production processes after 1870.

Appendix B

Survey of Ironworks in Southeastern New England, 1645 – 1848

Ironworks located to the north and west of Boston are in parentheses ()

This is a preliminary listing of the forges and furnaces of southeastern New England based on a review of numerous town histories and inquiries made to area town libraries and historical societies. This survey is not intended to be considered anything other than a rough sketch of colonial and early Republic ironworks; additional research by other individuals or institutions will hopefully enlarge our understanding of the diversity and complexity of early New England ironworking facilities and toolmakers. Corrections, suggestions, and additional listings are greatly appreciated.

Town	Location/Name	Dates	Process type	Owner	Sources*
Rowley (now Ipswich)		1636	Forge	Edmund Bridges	Mulholland
Saugus	Saugus River/ Hammersmith (Saugus Ironworks)	1644 or 1645 - 1688	Integrated ironworks (blast furnace, finery, chafery)	Richard Leader also James and Henry Leonard	Hartley, Ingram, Mulholland
Braintree	Monatiquot River	1645 - 1646	Forge and finery	John Winthrop Jr. & Richard Leader	Old Colony, Galer, Mulholland
Quincy	Furnace Brook	1646	Blast furnace	Henry Leonard	Galer, Hartley
Saugus	Saugus River/ Hammersmith (Saugus Ironworks)	1647 - 1652	Forge for scythe and edge tool production	Joseph Jenks (later Jencks)	Ingram, Hartley, Mulholland
Pembroke		1648 - 1650 ?	Iron	Timothy Hatherly	Briggs
Hanover	Drinkwater River	+/-1650	Forge, burned in 1676 and then rebuilt	Magoon and Wade	Briggs
Taunton	Two Mile River	1652 (1656) - 1777	Bloomery	Henry and James Leonard (d. 1691), Thomas Leonard (d.1713)	Old Colony, Mulholland

Town	Location/Name	Dates	Process type	Owner	Sources*
Providence	Blackstone River	1655	Forge	Joshua Foote	Bining
Dartmouth	Russells Mill	1656 - 1660	Bloomery	Ralph Russell	Ingram
(Concord)		1660 - 1682	Bloomery	Simon Lynke and Thomas Brattle	Mulholland
Concord	Assabet River at Westfield	1660 - 1705	Forge	Oliver Purchas	Hudson
Bourne	Back River Aptucxet	1662 or earlier	Forge	Ezra Perry	
(Rowley Village) (Boxford)		1668	Bloomery	Henry and Nathaniel Leonard (son)	Hurd, Mulholland
Taunton	Mill River	1670 - +/-1780	Bloomery	James Leonard & 3 sons (Josiah, Ben, Uriah)	Old Colony
Pawtucket, RI	Blackstone River	1671 - 1675	Forge	Joseph Jencks (son of Joseph Jenks)	Old Colony
Kingston	Hall's Brook	+/-1680 – 1734	Bloomery		Elliott
Bourne	Manomet River?	1681	Forge	John Pinne	Sawyer
Wareham	Agawam River	1685-1795	Mill	Israel Fearing	Chatier
Marion (Old Rochester)	Sippican River / Leonard's Forge	1690?	Bloomery	? Leonard	Leonard
Middleboro	Nemasket River & Tispequin Pond / Fall Brook Forge	1692 - 1735	Forge, enlarged in 1735	Unknown	Weston
North Dighton	Three Mile River	1695 enlarged after 1739	Bloomery	Richard Stevens	Old Colony

Town	Location/Name	Dates	Process type	Owner	Sources*
Dighton	Taunton River	1697 - 1708	Shipsmith	Robert Crossmen	Old Colony
Norton	Stony Brook / Chartley Ironworks	1698 - 1713	Bloomery	Thomas Leonard	Old Colony, Ingram
Bridgewater		1700 +	Integrated ironworks		Old Colony
Pembroke	Herring Pond Furnace Pond	1702	Blast furnace / casting	Lambart Despard	Briggs
Hingham	Ganit's River Above Pratt's Mill	1703?	Forge		(Hingham)
Hingham	Cohasset / Pratt's Mill	1703	Forge	Thomas Andrews, Aaron Pratt, Thomas James, etc.	(Hingham)
Abbington	Lower Beech Hill Brook	1704	Iron forge	Isaac Hobart	Whitman PL
Abbington	Beech Hill Brook	1704	Iron forge (bog iron from Great Pond, Weymouth)	Theodosuis Moore	Whitman PL
Pembroke	Indian Head River at Luddams Ford	1704 - 1735	Anchor forge and shipsmith	Thomas Bardin	Briggs
Bridgewater	Town River at Carver's Pond	1707	Forge	David Perkins	Bridgewater PL
Whitman	Satucket Path	1709	Forge, blacksmith	John Harden	Whitman PL
Bourne	Back River	-1710 - 1732	Ironworks	Deacon Alijah Perry	
Hanover	Drinkwater River / Mighills Iron Works	1710 +	Forge: cannons and anchors	Mighill	Briggs
Concord	Forge Pond Ironworks	1715	Forge		Hudson
N. Easton	Trout Hole Brook / Brummagen Forge	1716 - 1720, (1720 - 1742?)	Forge	Capt. James Leonard & son Eliphalet	Chaffin

Town	Location/Name	Dates	Process type	Owner	Sources*
Hanover	Indian River	1720 - 1795	Forge and furnace molded cannon balls	Captain Joseph Barstow (d. 1728) + sons	Briggs
S. Easton	Stone Pond / Leonard's Forge	1720	Forge	Capt. James Leonard	Chaffin
Easton	Cranberry Meadow	1724	Forge	Timothy Cooper	Chaffin
Taunton	Littleworth Brook / Kings Furnace	1724 - 1824	Blast furnace, (hollowware) rebuilt in 1816	John King, Elkanah Leonard	Old Colony, Nowak
Cumberland, RI	Cumberland Hill	+/-1725	Blast furnace – used rock ore		Old Colony
Dartmouth	Bedford Village	1725?	Forge	Elnanthan Sampson	Ellis
W. Bridgewater	Town River	1730	Blacksmith forge	Thomas Ames (son is James)	Galer
Plympton (S. Carver)	Weweantic River at Watson Cove Brook / Pope's Point	1732 - 1830	Blast furnace; first furnace to cast "Carver" hollowware	Isaac Lathrope, etc. (see Griffith 1913, *History of Carver*)	Murdock, Griffith
Kingston	Pine Brook	1734 - 1830	Forge, became a slitting mill in 1800	William Cook and Samuel Seabury	Elliott
Kingston	Trout Brook / Faunce's Furnace	1735	Blast furnace, made hollowware	Faunce	Elliott
Middleboro	Fall Brook	1735	Blast furnace	Peter Bennet, Francis Miller	Weston

Town	Location/Name	Dates	Process type	Owner	Sources*
Pembroke	Indian Head River at Luddams Ford	+/-1735 - 1774	Forge	Captain Thomas Bardwin (died 2/14/1774)	Briggs
N. Middleboro	Nemasket River at Muttock / Oliver's Furnace	+/-1735 - 1778	Blast furnace, forge, & slitting mill: cannon, mortars, howitzers, shot, & shell	Peter Oliver	Weston
Taunton	Mill River / Hopewell Ironworks	1739	Bloomery	Capt. Zephaniah Leonard	Old Colony
Bridgewater	Town River / Orr Forge	1740 - 1748 (1760)	Forge: edge tools, muskets, whitesmith & shipsmith	Hugh Orr	Bridgewater PL, Galer, (DATM)
N. Easton	Trout Hole Brook / Brummagen Forge	1742 - 1782	Forge	Eliphalet Leonard (son is Jacob)	Chaffin
Abbington	Beech Hill Brook / Hobart Bell Foundry	1745 - 1759	Foundry: bells, etc.	Isaac Hobart	Whitman PL
South Middleboro	Stillwater Stream / Stillwater Furnace	+/- 1750	Blast furnace: hollowware	Capt. Zenias Wood	Weston
N. Middleboro	Titicut River / Squawbetty Forge	+/- 1750	Ironworks		Weston
Plympton	Winnetuxet River	+/-1750 - 1869	Forge & smithy (later shovel forge)	Nathaniel Thomas (Jonathan Parker)	Plympton HS map

Town	Location/Name	Dates	Process type	Owner	Sources*
Hanover	Drinkwater River	1750 rebuilt 1830	Blast furnace: cast bells, stoves, hollowware, later machinery	Mr. Sylvester	Briggs
Plympton	Dennetts Pond at the Jones River	1750	Cannon foundry (nail factory 1823 – 1890)	Gideon Brand (Ebenezer Lobdell)	Plympton HS map
Easton	Beaver Brook / Furnace Village	1751 - 1798	Blast furnace and forges	John Williams then James Perry	Chaffin
Abbington (Whitman)	Beech Hill Brook /	1759	Blast furnace, 2 forges, made cannonballs, hollowware, later cannon for Rev. War, first air furnace, use of mountain ore	Isaac and Aaron Hobart	Whitman PL, Ingram
Bridgewater	Matfield River / Masschusetts State Furnace	1760 - 1775	Blast furnace, later made cannon for Rev. army	Col. Hugh Orr	Galer, Mitchell
Bridgewater	Matfield River	1760	Slitting mill	David & George Keith	Bridgewater PL
South Carver	Sampson's Pond / Charlotte Furnace	1760 - 1905	Blast furnace	Col. Bartlett Murdock, Benjamin Ellis	Murdock, Griffith
Easton	Marshall Place	1765 f.	Forge	Eliphalet Leonard	Chaffin
Middleboro	Near Mt. Carmel / Waterville Furnace	+/- 1770 - 1840	Blast furnace: hollowware	Nathaniel and William S--ddy	Weston
Bridgewater	Town River	1773	Forge: shovels, gunsmith during Revolution	Capt. John Ames (son Oliver, b. 1779)	Bridgewater PL, Galer

Town	Location/Name	Dates	Process type	Owner	Sources*
Easton	Marshall Place	1775 f.	2 steel furnaces added by 1775-76	Eliphalet Leonard & son Jonathan Leonard	Chaffin
Bridgewater	Town River / Orr Forge	1775 - 1783	Forge: gunsmith	Col. Hugh Orr, son is Robert Orr	Mitchell, Galer, Ingram
Bridgewater	Matfield River / North Furnace Mill	1775 - 1783	Furnace and Forge: bayonets, cannons	Col. Hugh Orr	Mitchell, Galer, Ingram
Bridgewater	Town River	+/- 1776	Forge: scythes & shovels, to Springfield Arsenal by 1804	Robert Orr	DATM
Bridgewater	Town River	1776	Furnace	Jeremiah Keith	Bridgewater PL
Taunton	Mill River / Hopewell Ironworks	1776	Hopewell Ironworks became a rolling and slitting mill	John Adams	Old Colony
Middleboro	Nemasket River	1776 +	Forge, foundry, & 2 finishing shops: shovel factory	Gen. Able Washburn & Thomas Weston	Weston
Pembroke	Indian Head River at Luddams Ford	1778 - 1791	Forge: anchors and shipsmiths	Josselyn family, w/ Lemuel Curtis then Reuben Curtis partners	Briggs
Pembroke	Indian Head River at Luddams Ford	1779 - Civil War	Anchors	Curtis' Forge, Ruben Curtis and Curtis family	Briggs

Town	Location/Name	Dates	Process type	Owner	Sources*
Kingston	Mile Brook	+/-1780	Shovel works	John Adams	Elliott
Middleboro	Pleasant St	+/- 1780	Blast furnace	Mr. Stafford	Weston
Mattapoisett	Mattapoisett River	1780 - 1860	Bloomery		Mendell
N. Easton	Trout Hole Brook / Brummagen Forge	1782 - 1800	Forge	Jacob Leonard (son is Isaac)	Chaffin
Bridgewater	Town River / Perkins Ironworks, later Bridgewater Iron Company	1785	Integrated ironworks, later made sheathing for USS Monitor (armor plating)	David Perkins & Nathaniel Lazell	Whitman PL, Galer
Easton	Marshall Place	1787 - 1808	Steel furnace	Jonathan Leonard	Chaffin
South Middleboro	Gt. Quittacus Pond / Stillwater Furnace	+/- 1790	Bloomery	Capt. Zenas Wood	Weston
Bourne		1790	Forge	Elisha Bourne	
Kingston	Jones River Trip Hammer	1791	Forge then anchors 1800 – 1852	Jedodiah Holmes, Jediah Holmes Jr.	Elliott
Easton	Shovel Shop Pond	1792 - 1801	Forge with a trip hammer, nailer's shop	Eliphalet Leonard III	Chaffin
Carver	Crane Brook / Federal Furnace	1793	Blast furnace		Murdock
Hanover	Drinkwater River	1794	2 forges: anchors	unknown	Briggs
Kingston	Trout Brook	1794	Anchor forge		Elliott
Hanover	Indian River / Barstow's Forge	1795 - 1837	Forge: wrought iron	Robert Salmon	Briggs

Town	Location/Name	Dates	Process type	Owner	Sources*
Walpole	Neponset River	1796 f.	Forge: farm implements, shovels?	Oliver Ames, employee	Galer
West Wareham / Tremont	Weweantic River / Little England	1798 - 1805	Forge and furnace	Four Leonard brothers	Chatier
Easton	Beaver Brook / Furnace Village	1798 - 1804	Blast furnace and forges	John Brown	Chaffin
N. Easton	Trout Hole Brook / Brummagen Forge	1800 - 1802	Forge	Isaac Leonard	Chaffin
Easton	Marshall Place	1800 - 1808	Steel furnaces	Jonathan Leonard & Eliphalet Leonard II	Chaffin
Carver	Weweantic River / Fresh Meadow Fur.	1800	Blast furnace		Murdock
Kingston	Halls Brook	1800 - 1881	Anchor forge		Elliott
Easton	Shovel Shop Pond	1801 - 1803	Forge with a trip hammer, nailer's shop	Daniel Wheaton, Ebenezer Alger	Chaffin
N. Easton	Queset River / Shovel Shop Pond	1803 - 1950	Forge, steel furnace, & shovel factory	Oliver Ames (d. 1863)	Galer, Chaffin
N. Easton	Hoe Shop Dam below Shovel Shop Pond	1804	Forge	Nathan Pratt	Chaffin
Easton	Beaver Brook / Furnace Village	1804 - 1823	Blast furnace and forges	Cyrus Alger	Chaffin
Kingston	Stony Brook	1805	Ironworks (shipsmith)	Seth Washburn, Deacon Seth Drew	Elliott

Town	Location/Name	Dates	Process type	Owner	Sources*
Wareham / Tremont	Weweantic River at Horseshoe Pond	1805 - 19352	Iron mill later called Standard Horseshoe Works	Four Leonard brothers	Chatier
Kingston	Pall Mall	< 1808	Forge and trip hammer		Elliott
Kingston	Trout Brook	1810	Forge (spades and shovels)	John and Eleazer Faunce	Elliott
Falmouth	Quissett Harbor, later Eel Pond, Woods Hole	+/-1811 - 1873	Shipsmith	Braddock Gifford	Smith (1983)
Wareham	Wankinco River / Tremont Nail Co.	1812 f.	Nails		Chatier
Canton		1813 - 1827 f.	Steel furnace	Leonard & Kinsley (Amos Binney after 1827)	Chaffin
Kingston	Smelt Brook	1815	Ironworks (auger maker) later Old Colony Foundry	Thomas Cushman	Elliott
Hanover	Drinkwater River	1816	Forge: bar iron and anchors	Bates and Holmes	Briggs
Wareham	I&J Pratt and Co. later Wareham Iron Co.	1819 - 1834	Bloomery	Jarad Pratt	Weston, Chartier
Carver	Wancinko Swamp / Slug Pond	1819	Blast furnace	Lewis Pratt	Murdock
West Wareham	Mt. Washington Ironworks later Tremont Ironworks	+/-1820	Puddling furnace & rolling mill	Col. Bartlett Murdock	Wareham PL

Town	Location/Name	Dates	Process type	Owner	Sources*
Wareham	Agawam River / Agawam Iron Mill	1820 - 1836	Hollowware	Thomas Saver	Chatier
Hanover	Longwater Stream / Brook's Mill	1820 - 1850	Cast iron plows		Briggs
Plympton	Winnetuxet River	1822	Rolling mill: nails & bolts	Oliver Perkins	Plympton HS map
Bourne	Pocasset River	1822	Cold blast cupola furnace	Hercules Weston	
Wareham / Tremont	Weweantic River	1822 - 1833 - 1845	Ironworks	Bartlett Murdock then Washington Steel	Chatier
Easton	Beaver Brook / Furnace Village	1823 - 1832	7 furnaces	Shepard Leader	Chaffin
Hingham	Weir River at Thomas Pond	1824 - 1828	Blast furnace: hollowware	Benjamin Thomas	(Hingham)
Carver	South Meadows	1824	Blast furnace	Benjamin Ward, Lewis Pratt	Murdock
Carver	Sampson Pond / Barrow's Furnace	1825	Blast furnace	Joseph and Nelson Barrows	Murdock
Carver	Wenham Brook / Wenham Blast Furnace	1827	Blast furnace, later became cupola furnace & foundry	Lewis Pratt	Murdock
Wareham	Wankinco River / Tihonet Works	1828	Puddling furnace, nail machines		Chatier
Wareham	Wareham Iron Company	1829	Bloomery (first use of Reed nail making machine)	Jarad Pratt	Weston
Wareham	Assawampsett Pond / Parker Mills	1829 - 1834	Cupola furnace	Isaac Pratt	Weston

Town	Location/Name	Dates	Process type	Owner	Sources*
Sagamore (Bourne)	Herring River	1829	Forge	Isaac Keith later Keith & Ryder	
Wareham	Wareham River	1830	Cupola furnace	Col. Bartlett Murdock	Chatier
Wareham	Weweantic River	1830 - 1831	Rolling mill, nail factory	Bartlett Murdock, George Howland	Chatier
Wareham	Weweantic River / Weweantic Iron Works	1831 - 1838 - 1860	Rolling mill, nail factory	J.B. Tobey later Lewis Kinney and Company	Chatier
Bourne	Pocasset River / Pocasset Foundry	1832	Blast furnace made hollowware	Kendrick McGraw	
Wareham	Wankinco River / Parker Mills	1834	Ironworks	John Avery Parker, William Rodham, Charles W. Morgan	Chatier
Wareham	Agawam River / Agawam Nail Company	1836 - 1845	Nails	Samuel Tisdale	Chatier
S. Hingham	Main St / Hammersmith	1836 - 1883	Hatchets and edge tools (steam power 1846)	Joseph Jacobs to Underhill in 1883	(Hingham)
Easton	Beaver Brook / Furnace Village	1837	Malleable ironworks	Daniel Belcher	Chaffin
South Hanover	Sylvester Forge / Hanover Forge	1837 - 1853	Forge	E. Y. Perry	Briggs
Kingston	Stony Brook	1837 f.	Toolmakers	C. Drew & Co.	Elliott
Hingham	Weir River? / Hingham Malleable Iron Company	1840	Cupola furnace, malleable ironware		(Hingham)

Town	Location/Name	Dates	Process type	Owner	Sources*
Hingham	?	+/-1840	Shipsmith (wrought iron spikes)	William Thomas	(Hingham)
Hingham	Summer St at harbor / Eagle Iron Foundry	1844 - 1846	Cupola furnace, burned in 1846		(Hingham)
Wareham	Agawam River at Glen Charlie	1845	Rolling mill	Samuel Tisdale	Chatier
Wareham / Tremont	Weweantic River / Tremont Iron Works	1845 - 1858	Ironworks	Eventually sold to Tremont Nail Co.	Chatier
Hingham	Accord Pond	1845 - 1870	Ax factory	Charles Whitney	(Hingham)
Kingston	Smelt Brook	1848 - 1987	Stoves & hollowware, later rivets & tacks	Cobb & Drew	Elliott
Bournedale	Manomet River	1849	Forge / edge toolmaker	Lewis Howe	
Bournedale	Manomet River	1850?	Forge / Ax maker	Seth Holloway	
Wareham	Weweantic River / Weweantic Iron Works	1860 - 1884	Rolling mill, nail factory	Leased to Tremont Nail Co.	Chatier
Wareham	Wareham River / Franconia Iron and Steel Co.	1864 - 1879			Chatier

*Sources: Bining 1933, Bridgewater Public Library personal communications, Briggs 1889, Chaffin 1886, Chatier 2007, DATM: Nelson 1999, Elliott 2005, Ellis 1892, Galer 2002, Griffith 1913, Hartley 1957, (History of Hingham), Hudson 1904, Hurd 1888, Leonard 1907a, Ingram 2000, Mendell 2004, Mitchell 1970, Mulholland 1981, Murdock 1937, Nowak, Old Colony Historical Society 1901, Plympton Historical Society map, Sawyer 1988, Smith 1983, Weston 1906, Wareham Public Library personal communications, Whitman Public Library personal communications.

Appendix C

Bining's listing of forges and furnaces located in the Province of Massachusetts Bay in 1758 (Bining 1933, 126-127)

Many of these ironworking facilities are not included in Appendix B.

A

A LIST OF FORGES AND FURNACES WITHIN THE PROVINCE OF THE MASSACHUSETTS BAY AND OF THE PERSONS' NAMES BY WHOM THEY ARE OWNED OR OCCUPIED. (1758)*

Counties	Towns	Forges	Furnaces	Owners or Occupiers
Suffolk	Dedham	1	–	William Avery
	Walpole	1	–	Morse
	Walpole	1	–	Capt. Clap
	Stoughton	1	–	Nathaniel Leonard
	Stoughton	1	–	Elkanah Billings
	Wrentham	–	1	Josiah Maxey
	Needham	1	–	Oliver Pratt
Essex	Salisbury	1	–	Philip Rowell
Middlesex	Westford	1	–	Keep
Hampshire	Springfield	1	–	Joseph Dwight
	Sheffield	1	–	John Ashley
Worcester	Western	2	–	Keys, Blackmore & Hayward
	Western	–	1	David Keys
	Leicester	1	–	Lieut. Tucker & Others
	Sutton	1	–	John Hazleton
	Uxbridge	1	–	Nicholas Baylis
Plymouth	Middleboro	1	–	Peter Oliver
	Middleboro	1	–	Elkanah Leonard
	Wareham	1	–	Mrs. Lothrop
	Rochester	1	–	Amos Bates
	Plympton	-	–	Warden & Goodwin
	Plympton	-	–	John Beccham

* This list was compiled by Andrew Oliver in 1758 for Governor Thomas Pownall, who sent it to the Board of Trade. Colonial Office Papers 5:889.

126

Counties	Towns	Forges	Furnaces	Owners or Occupiers
	Bridgwater	1	–	Capt Bass
	Bridgwater	1	–	Codman
	Bridgwater	1	–	Iona Carver
	Bridgwater	1	–	Daniel Hayward & Co.
	Bridgwater	1	un-improved	John Miller
	Kingston	1	–	Samuel Seabury
	Kingston	1	–	Jonathan Holmes
	Hanover	1	–	Thomas Jocelyn
	Pembroke	1	–	Elijah Cushing
	Hanover	1	–	Palmer
	Plympton	–	1	Joseph Scott & Co.
	Plympton	–	1	Mrs. Lothrop & Co.
	Kingston	–	1	Gamaliel Badford & Co.
	Halifax	–	1	Thos. Croade & Co.
	Bridgwater	–	1	Josiah Edson & Co.
	Bridgwater	–	1	Nicholas Sever & Co.
	Bridgwater	–	1	James Bowdoin
	Bridgwater	–	1	unimproved
	Hanover	–	1	Thomas Jocelyn
Barnstable	———	None		———
Bristol	Norton	1	–	George Leonard
	Norton	1	–	Ephraim Leonard
	Norton	1	–	William Stone
	Raynham	1	–	Elijah Leonard
	Taunton	1	–	Zephary Leonard
	Taunton	1	–	James Leonard
	Taunton	1	–	Baylis & Laughton
	Easton	1	–	Eliphalet Leonard
	Freetown	1	–	Ebenezer Hathaway
	Attleboro	1	–	Sweet & Merrit
	Taunton	–	1	Mrs. King
	Raynham	–	1	Seth Leonard
	Easton	–	1	Mrs. Williams
York	———	None		———
Dukes	———	None		———
Nantuckett	———	None		———
		41	14	

Appendix D

Listing of Shipwright Tools: 1607 - 1840

The basic woodworking tool kits of early New England settlers remained almost unchanged from the early years of the colonial period until the beginning of the 19th century. What did change was how the tools were made, who made them, and where they were made. New England's first shipbuilders often had access to only a limited selection of woodworking tools, the most important ones are shown in **bold** in the table below. As shipbuilding became more sophisticated in the 19th century, with larger ships of more complex design being constructed in or near communities with larger populations and toolmaking capabilities, any of the following tools could have been in a woodworkers' tool kit. After 1840, the advent of steam power, the rotary saw mill, and cast iron machinery gradually supplanted the function of many ships carpenter's and other woodworker's hand tools.

Later tools (post 1750) are enclosed in parentheses ( )

Axes	Adzes	Other Edge Tools	Other Hand Tools
felling	**lipped**	**gouge**	**caulking iron**
broad	peen (peg poll)	forming chisel	brace
mast	block (poll)	**mortising chisel**	bow drill
hewing	coopers	**corner chisel**	pump drill
mortising	mortising (gutter) adz	**framing chisel**	spud
lathing hatchet	**Measuring Tools**	turning tools	pickaroon
shingle hatchet	**dividers**	**slick**	log dog
coopers	calipers	pod auger (screw auger)	cant hook
twybil (Penn Dutch)	square (framing, try, etc.)	spoke shave	pike
ice ax	level	**draw knife**	peavey
Saws	chalk, line and reel	**mast shave**	grappling hook
pit	marking gauge	cooper's shave	ring dogs
frame	traveler	wedge	barking iron

whip	plumb bob	scorp	saw set
buck (bow)	**Hammers and Mallets**	froe	shaving horse
cross cut	wood mallet	gimlet	block and tackle
hand saw	beetle	timber scribe	trowel
tenon saw	**caulking mallet**	screw tap	screw clamp
fret saw	claw hammer	block knife	(breast drill)
Planes		chamfer knife	(mortising drill)
block			
smooth			
fore (or trying)		**Other Trades**	
moulding or molding	**Blacksmith**	**Cobbler**	**Farrier**
joining	**Currier**	**Flax dresser**	**Fuller**
plough or plow	**Weaver**	**Carriage maker**	**Lime burner**
mitre			
shoulder			
squaring			

Appendix E

Excerpts from Moxon

The following is an excerpt from Joseph Moxon's 1703 *Mechanick exercises or the doctrine of handiworks* (Moxon [1703] 1989, 56-62).

Moxon's excerpt is an excellent summary of an educated layperson's take on steel- and iron-making strategies and sources. Moxon makes notable references to forging wrought iron, case hardening, and the excellence of steel made in the Forest of Dean, later the location of R. F. Mushet's ironworks, that enabler of Bessemer's pneumatic bulk steel process. He also notes Danzig as a source of Swedish steel, cementation furnaces having been built there in the second half of the 17th century. He comments on the Spanish steel of Biscayne, used for French trading axes, German steel and the role of the Rhine River in its transport to German toolmaking centers, and the high quality of Venetian steel. He also notes both the excellence and difficulty of creating Damascus steel. Moxon continues with a survey of contemporary interpretations of annealing, hardening, and tempering steel and ends with a famous aphorism of the time: "form thick and ground thin."

to ſtop the Socket from ſliding too far upon
the *Shank*. From this Shoulder, the reſt of
the *Shank* muſt run Tapering down, to the
ſmall end the *Bullet-bore* is faſtned to. You
muſt Work with it, as you were taught to
Work with the *Square-bore*.

Of Twiſting of the Iron.

SQuare and flat Bars, ſometimes are by Smiths,
Twiſted for Ornament. It is very eaſily done;
for after the Bar is Square or flat Forg'd (and if
the curioſity of your Work require it truly Fil'd)
you muſt take a *Flame-heat*, or if your Work be
ſmall, but *Blood-red heat*, and you may twiſt it
about, as much or as little as you pleaſe, either
with the *Tongs*, *Vice* or *Hand-vice*, &c.

Of Caſe-hardning.

CAſe-hardning is ſometimes us'd by *File-cutters*,
when they make courſe *Files* for Cheapneſs,
and generally moſt *Raſps* have formerly been made
of Iron and *Caſe-hardned*, becauſe it makes the
outſide of them hard. It is us'd alſo by *Gun-
ſmiths*, for Hardning their Barrels; and it is
us'd for *Tobacco-boxes*, *Cod-piece-buttons*, *Heads* for
Walking-ſtaves, &c. And in theſe Caſes, Workmen to ſet a greater value on them in the Buyers
eſteem, call them *Steel-barrels*, *Steel-tobacco-boxes*,
Steel-buttons, *Steel-heads*, &c. But Iron thus
hardned takes a better Poliſh and keeps the Poliſh much longer and better, than if the Iron
were not *Caſe hardned*. The manner of *Caſe-
hardning* is thus, Take *Cow-horn* or *Hoof*, dry it
thoroughly in an Oven, and then beat it to Powder, put about the ſame quantity of Bay-Salt to it,
and mingle them together with ſtale Chamberly,
or elſe White-wine-vinegar. Lay ſome of this
mixture upon the Loam, made as you were
taught

taught *Numb. I. fol.* 13. And cover your Iron all over with it; then wrap the Loam about all, and lay it upon the Hearth of the Forge to dry and harden: When it is dry and hard, put it into the Fire and blow up the Coals to it, till the whole Lump have juft a *Blood-red-beat*, but no higher, left the quality of your mixture burn away and leave the Iron as foft as at firft. Then take it out and quench it: Or, inftead of Loam, you may wrap it up in Plate Iron, fo as the mixture may touch every part of your Work, and blow the Coals to it, as aforefaid.

Of feveral forts of Steel in common ufe among Smiths.

THE difficulty of getting good Steel makes many Workmen (when by good hap they light on it) commend that Country-Steel for beft, from whence that Steel came. Thus I have found fome cry up *Flemifh-fteel*, others *Swedifh, Englifh, Spanifh, Venice*, &c. But according to my Obfervation and common Confent of the moft ingenious Workmen, each Country produces almoft indifferently good and bad; yet each Country doth not equally produce fuch Steel, as is fit for every particular purpofe, as I fhall fhew you by and by. But the feveral forts of Steel, that are in general ufe here in *England*, are the *Englifh*, the *Flemifh*, the *Swedifh*, the *Spanifh* and the *Venice-fteel*.

The *Englifh-fteel* is made in feveral places in *England*, as in *Yorkfhire, Gloucefterfhire, Suffex*, the *Wild of Kent*, &c. But the beft is made about the *Forreft of Dean*, it breaks Fiery, with fomewhat a courfe Grain But if it be well wrought and proves found, it makes good Edge-tools, Files and Punches. It will work well at the Forge, and take a good Heat.

The

E-3

The *Flemish-steel* is made in *Germany*, in the Country of *Stiermark* and in the *Land of Luyck*: From thence brought to *Colen*, and is brought down the River *Rhine* to *Dort*, and other parts of *Holland* and *Flanders*, some in *Bars* and some in *Gads*, and is therefore by us call'd *Flemish-steel*, and sometimes *Gad-steel*. It is a tough sort of Steel, and the only Steel us'd for Watch-springs. It is also good for Punches; File-cutters also use it to make their Chissels of, with which they cut their Files. It breaks with a fine Grain, works well at the Forge, and will take a welding Heat.

I cannot learn that any Steel comes from *Sweden*, but from *Dantzick* comes some which is call'd *Swedish-steel*: It is much of the same Quality and Finess with *Flemish-steel*.

The *Spanish-steel* is made about *Biscay*. It is a fine sort of Steel, but some of it is very difficult to work at the Forge, because it will not take a good Heat; and it sometimes proves very unsound, as not being well *curried*, that is well wrought. It is too quick (as Workmen call it) that is, too brittle for Springs or Punches, but makes good fine Edg'd-tools.

Venice-steel is much like *Spanish-steel*, but much finer, and Works somewhat better at the Forge. It is us'd for Razors, Chirurgion's Instruments, Gravers, &c. Because it will come to a fine and thin Edge. Razor makers generally clap a small Bar of *Venice-steel* between two small Bars of *Flemish-steel*, and so Work or Weld them together, to strengthen the back of the Razor, and keep it from cracking.

E-4

There

There is another fort of Steel, of higher commendations than any of the forgoing forts. It is call'd *Damafcus-fteel*; 'tis very rare that any comes into *England* unwrought, but the *Turkifh-Cymeters* are generally made of it. It is moft difficult of any Steel to Work at the Forge, for you fhall fcarce be able to ftrike upon a Bloodheat, but it will *Red-fear*; infomuch that thefe *Cymeters* are, by many Workmen, thought to be caft Steel. But when it is wrought, it takes the fineft and keeps the ftrongeft Edge of any other Steel. Workmen fet almoft an ineftimable value upon it to make Punches, Cold-punches, &c. of. We cannot learn where it is made, and yet as I am inform'd, the Honourable Mr. *Boyl* hath been very careful and induftrious in that enquiry; giving it in particular charge to fome Travellers to *Damafcus*, to bring home an Account of it: But when they came thither they heard of none made there, but were fent about 50 Miles into the Country and then they were told about 50 Miles farther than that : So that no certain Account could be gain'd where it is made. *Kirman* towards the Ocean affords very fine Steel, of which they make Weapons highly priz'd ; for a *Cymeter* of that Steel, will cut through an Helmet with an eafie blow. *Geog. Rect. foi.* 279.

The Rule to know good Steel by.

BReak a little piece of the end of the Rod, and obferve how it breaks ; for good Steel breaks fhort of all Gray, like froft work Silver. But in the breaking of the bad you will find fome veins of Iron fhining and doubling in the Steel.

E-5

Of

Of Nealing of Steel.

HAving chofe your Steel and forg'd it to your intended fhape, if you are either to File Engrave or to Punch upon it, you ought to Neal it firft, becaufe it will make it fofter and confequently work eafier. The common way is to give it a *Blood-red-heat* in the Fire, then take it out, and let it cool of it felf.

There are fome pretenders to know how to make Steel as foft as Lead; but fo oft as my Curiofity has prompted me to try their pretended Proceffes, fo oft have they fail'd me ; and not only me, but fome others, careful Obfervers. But the way they moft boaft of, is the often heating the Iron or Steel in red-hot Lead, and letting it cool of it felf with the Lead. I have many times try'd this without any other fuccefs, than that it does make Iron or Steel as foft as if it were well Neal'd the common way, but no fofter : And could it be otherwife, the fmall Iron Ladles, that Letter-founders ufe to the cafting of Printing Letters, would be very foft indeed ; for their Iron Ladles are kept conftantly Month after Month in melting Mettal, whereof the main Body is Lead, and when they caft fmall Letters, they keep their Mettal redhot ; and I have known them many times left in the Mettal and cool with it, as the Fire has gone out of it felf; but yet the Iron Ladles have been no fofter, than if they had been well Neal'd the common way. But perhaps thefe Pretenders mean the Iron or Steel fhall be as foft as Lead, when the Iron or Steel is red-hot ; if fo, we may thank them for nothing.

But

But that which makes Steel a very small matter softer than the common way of Nealing is, by covering Steel with a courſe Powder of Cow-Horns, or Hoofs, or Rams-Horns, and ſo incloſing it in a Loam: Then put the whole Lump into a Wooden Fire to heat red-hot and let it lie in the Fire till the Fire go out of it ſelf, and the Steel cool with the Fire.

Of Hardning and Tempering Steel.

ENngliſh, *Flemiſh* and *Swediſh-ſteel*, muſt have a pretty high heat given them, and then ſuddenly quench in Water to make them very hard; but *Spaniſh* and *Venice-ſteel* will need but a Blood-red-heat, and then when they are quench'd in Water, will be very hard. If your Steel be too hard, that is to brittle, and it be an edg'd or pointed Inſtrument you make, the edge or point will be very ſubject to break; or if it be a Spring, it will not bow, but with the leaſt bending it will ſnap aſſunder: Therefore you muſt *let it down* (as Smiths ſay) that is, make it ſofter, by *tempering* it: The manner is thus, take a piece of Grin-ſtone or Whet-ſtone and rub hard upon your Work to take the black Scurf off it, and brighten it; then let it heat in the Fire, and as it grows hotter you will ſee the Colour change by degrees, coming to a light goldiſh Colour, then to a dark goldiſh Colour, and at laſt to a blew Colour; chooſe which of theſe Colours your Work requires, and then quench it ſuddenly in Water. The light goldiſh Colour is for Files, Cold-chiſſels and Punches, that Punch into Iron and Steel: The dark goldiſh Colour for Punches to uſe on Braſs, and generally for moſt Edge-tools: The blew Colour gives the Temper to Springs in general, and is alſo us'd to Beautifie both Iron and Steel; but then Workmen ſome-times

E-7

times grind *Indico* and *Sallad-oyl* together, and rub that mixture upon it, with a woollen Rag, while it is heating, and let it cool of it felf.

There is another fort of *Hardning*, call'd *Hammer-hardning*. It is moft us'd on Iron or Steel Plates, for *Dripping pans*, *Saws*, *Straight-Rulers*, &c. It is perform'd only, with well Hammering of the Plates, which both fmooths them, and beats the Mettal firmer into its own Body, and fomewhat hardens it.

The manner of Forging Steel, either for Edge-tools, Punches, Springs, *&c.* Is (the feveral fhapes confider'd) the fame with forging Iron: Only this general Rule obferve, from an old *Englifh* Verfe us'd among Smiths, when they Forge Edge-tools,

> *He that will a good Edge win,*
> *Muft Forge thick and Grind thin.*

The End of Smithing.

MECHA-

Appendix F

Chapelle's listing from *The National Watercraft Collection*

This listing of domestic built ships in the National Watercraft Collection is extracted from: Howard I. Chapelle, 1960, *The National Watercraft Collection*, United States National Museum Bulletin 219, Washington, DC: Smithsonian Institution. The surprising number of downeast-built ships listed in the watercraft collection, in contrast to the lack of surviving half models of Bath, Maine or Massachusetts-built ships (other than Newburyport), provides a startling insight into the extent by which Maine communities participated in the booming shipbuilding economy of mid-19th century Maine.

Abbreviations: C – coasting (coastwise trade); F – foreign trade (transoceanic); L – lumber; Sch – schooner; 2MSch – two-masted schooner; 3MSch – three-masted schooner; SQTPS – square topsail; WI – West Indies.

Rig	Trade	Ship Name	Location	Length	Date	Tonnage
Ship	Ocean freighter	Atticus	Castine	132	1818	298
Ship	Cotton	Mayflower	Bath	131	1855	
Ship	Ocean freighter	Oregon	Bath		1878	431
Bark	European	Crusade Europe	Milbridge	216	1854	
Bark	Ocean freighter	Julia	Ellsworth	164	1877	758
Brigantine	West Indies	unknown	Bath	91	1825	
Brigantine	West Indies	Amethyst	Sullivan	82	1838	
Brigantine	C/WI	Watson	Sedgwick	90	1846	146
Brigantine	Coasting	Telula	Cherryfield	78	1848	
Brigantine	C/WI/Lumber	Iscarion	Lamoine	89	1852	198
Brigantine	Coasting	Abby Watson	Sedgwick	109	1852	214
Brigantine	C/WI	unknown	Bath	90	1852	
Brigantine	West Indies	Fredonia	Ellsworth	103	1854	
Brigantine	West Indies	Anita Owen	Milbridge	117	1853	
Brigantine (2)	West Indies	Eva M. Johnson Mary E Punnel	Harrington	114	1868	
Brigantine (2)	C/WI	Antelope & Gazelle	Harrington	113	1866	319 326
Brigantine	C/F	Minnie Smith	Milbridge	116	1871	
Brigantine	C/WI	J W Parker	Belfast	129	1874	
SQTPS Schooner	C/WI	Ruth Thomas	Franklin	88	1845	

Rig	Trade	Ship Name	Location	Length	Date	Tonnage
SQTPS Schooner (2)	C/L	Arrowsic Eagle	Bath Arrowsic's	85	1847	
Sch	C/L	Watchman	Tinker's Island	80	1847	
Sch	C/WI	Marcia Trebow	Bucksport	89	1847	
Sch	C/L	Lucy	Sargentville	77	1852	
Sch	WI	Wakeag	Lamoine	102	1855	
Sch	Coasting	J. W. Hale	Brooklin	87	1855	
Sch (obsolete clusy on document)	C/L	North Star	Sullivan	60	1856	
Sch	C/WI	Aaron	Lubec	108	1858	
Sch	C/L	E Closson	Sedgwick	95	1860	135
Sch	C/L	Ada S Allen	Dennysville	98	1867	142
Sch	Coasting	Mountain Laurel	Lamoine	96	1868	141
Sch	C/L	Mable F Staples	Machias	124	1869 + 1871	268
Sch	Lumber	Alzema	Harrington			
2MSch (2)	Coasting	William H Ardus Lenora	Ellsworth	89	1821 1837	90
Sch	Coasting	D S Lawrence City of Ellsworth	Ellsworth	63	1875 1875	
Sch	Coasting	Bloomer	Eden	64	1855	51
Sch (3)	Coasting	Helen Alta V Cole Pojara	Harrington	119	1874-5	
Sch	Lime	Hunter	Orland	116	1876	187
3MSch (2)	Lumber	Nellie S Pickering Fame Gorham	Belfast	135	1870	
3MSch (2)	West Indies	Janie M Riley Susan P Thurlow	Harrington	133	1872	440
3MSch	C/WI/Lime	John Bird	Rockland	131	1872	
3MSch	C/WI	William F Frederick	Belfast	135	1874	
3MSch	Lime	Meyer & Muller	Belfast	163	1883	

Appendix G

Art of the Edge Tool Bibliography

The following bibliography consists of the sources used to compile the volumes in the *Hand Tools in History* series. A number of these citations appear in other volumes of this series. A more detailed special topic bibliography on ferrous metallurgy is included in volume 11, *Handbook for Ironmongers*. Numerous additional citations pertaining to New England's maritime and industrial history and American steel- and toolmaking strategies and techniques appear in the preceding and following volumes of this series.

Alexander, Jennie and Follansbee, Peter. 2012. *Make a joint stool from a tree: An introduction to 17th-century* joinery. Ft. Mitchell, KY: Lost Art Press.

Anonymous. He could outrun Webster: Jason Murdoch's queer encounter with the statesman when gunning for peep – is Wareham's oldest citizen. *Boston Globe.*

Abell, Sir Westcott. 1948. *The shipwright's trade.* Cambridge: The University Press.

Albion, Robert Greenhalgh, William A. Baker, and Benjamin Woods Labaree. 1972. *New England and the sea.* Middletown, CT: Wesleyan University Press.

Allen, Richard Sanders. 1981. *Iron works in Bristol county, Massachusetts.* Unpublished.

American History Department of Jonesport-Beals High School. 1970a. *A pictorial history of the town of Beals, Maine.* Calais, ME: Advertiser Publishing Company.

American History Department of Jonesport-Beals High School. 1970b. *A pictorial history of the town of Jonesport, Maine.* Calais, ME: Advertiser Publishing Company.

Andrews, Charles M. 1934. *The colonial period in American History.* 4 vols. New Haven, CT: Yale University Press.

Aston, James, and Edward B. Story. 1939. *Wrought iron: Its manufacture, characteristics and applications.* Pittsburgh, PA: A. M. Byers Company.

Atchison, Mark. (September 2012). William Palmer–An English nailmaker in New England. *The Chronicle.* 65(3). pg. 89.

Bailey, Sarah Y. 1926. *The civic progress of Kingston*. Kingston, MA: Two Hundredth Anniversary Committee of the Town of Kingston.

Bailyn, Bernard. 1964. *The New England merchants in the seventeenth century*. New York: Harper Torch Books.

Bailyn, Bernard, and Lotte Bailyn. 1959. *Massachusetts shipping, 1697-1714: A statistical study*. Cambridge: Harvard University Press.

Baker, William A. 1966. *Sloops & shallops*. Barre, MA: Barre Publishing Co.

Baker, William A. 1973. *A maritime history of Bath, Maine and the Kennebec region*. 2 vols. Bath, ME: Maritime Research Society of Bath.

Banks, Charles Edward. 1966. *The history of Martha's Vineyard Dukes County Massachusetts in three volumes: Volume II town annals*. Edgartown, MA: Dukes County Historical Society.

Barlow, Raymond E., and Joan E. Kaiser. 1985. *The glass industry in Sandwich*. vol. 3. Atglen, PA: Schiffer Publishing Ltd.

Barraclough, K.C. 1984a. *Steelmaking before Bessemer: Blister steel, the birth of an industry*. Volume 1. London: The Metals Society.

Barraclough, K.C. 1984b. *Steelmaking before Bessemer: Crucible steel, the growth of technology*. Volume 2. London: The Metals Society.

Barry, John S. 1855-57. *The history of Massachusetts*. Boston: Phillips, Sampson and Company.

Bates, Charles L. Brief history of the town of Wareham. Chapter from unknown text. 109-117.

Bining, Arthur Cecil. 1933. *British regulation of the colonial iron industry*. Philadelphia: University of Pennsylvania Press.

Blodgette, George Brainard. 1933. *Early settlers of Rowley, Massachusetts*. Rowley, MA: Amos Everett Jewett.

Bober, Harry. 1981. *Jan van Vliet's book of crafts and trades: With a reappraisal of his etchings*. Albany: Early American Industries Association.

Bolles, Albert S. 1878. *Industrial history of the United States*. Norwich, CT: Henry Bill Publishing Co.

Boucher, Jack E. 1964. *Of Batsto and bog iron*. Batsto, NJ: The Batsto Citizens Committee.

Brack, H. G. 2008. *Hand Tools in History*. 6 vols. Hulls Cove, ME: Pennywheel Press.

Brack, H. G. 2010. *Registry of Maine Toolmakers*. Hulls Cove, ME: Pennywheel Press.

Bradford, John. 2007. Designing a 17th century pinnace. Paper presented at the 35th Albert Reed and Thelma Walker Maritime History Symposium, May 4-6, in Bath, Maine.

Bradford, John. 2011. *The 1607 Popham Colony's pinnace Virginia: An in-context design of Maine's first ship*. Rockland, ME: Maine Authors Publishing.

Bradford, Gershom. 1954. The Ezra Westons, shipbuilders of Duxbury. *American Neptune* 14: 29-41.

Brady, William. [1847] 2002. *The kedge-anchor; or, young sailors' assistant*. Mineola, NY: Dover Publications Inc.

Brain, Jeffrey Phipps. 1995. *Fort St. George: Archaeological investigation of the 1607-1608 Popham colony on the Kennebec*. Salem, MA: Peabody Essex Museum.

Brain, Jeffrey Phipps. 1997. *Fort St. George. II: Continuing investigation of the 1607-1608 Popham colony on the Kennebec River in Maine*. Salem, MA: Peabody Essex Museum.

Brain, Jeffrey Phipps. 1998. *Fort St. George III: 1998 excavations at the site of the 1607-1608 Popham colony on the Kennebec River in Maine*. Salem, MA: Peabody Essex Museum.

Brain, Jeffrey Phipps. 1999. *Fort St. George IV: 1999 excavations at the site of the 1607-1608 Popham colony on the Kennebec River in Maine*. Salem, MA: Peabody Essex Museum.

Brain, Jeffrey Phipps. 2001. *Fort St. George VI: 2001 excavations at the site of the 1607-1608 Popham colony on the Kennebec River in Maine*. Salem, MA: Peabody Essex Museum.

Brain, Jeffrey Phipps. 2002. *Fort St. George VII: 2002 excavations at the site of the 1607-1608 Popham colony on the Kennebec River in Maine.* Salem, MA: Peabody Essex Museum.

Brain, Jeffrey Phipps. 2003. *Fort St. George. VIII: 2003 excavations at the site of the 1607-1608 Popham colony on the Kennebec River in Maine.* Salem, MA: Peabody Essex Museum.

Brain, Jeffrey Phipps. 2004. *Fort St. George. IX: 2004 excavations at the site of the 1607-1608 Popham colony on the Kennebec River in Maine.* Salem, MA: Peabody Essex Museum.

Brain, Jeffrey Phipps. 2005. *Fort St. George. X: 2005 excavations at the site of the 1607-1608 Popham colony on the Kennebec River in Maine.* Salem, MA: Peabody Essex Museum.

Brain, Jeffrey Phipps. 2007a. *Fort St. George: Archaeological investigation of the 1607-1608 Popham colony.* Occasional Publications in Maine Archaeology. Number 12. Augusta, ME: The Maine State Museum, The Maine Historic Preservation Commission, and The Maine Archaeological Society.

Brain, Jeffrey Phipps. 2007b. *Fort St. George. XI: Excavations at the site of 1607-1608 Popham colony on the Kennebec River in Maine.* Salem, MA: Peabody Essex Museum.

Brain, Jeffrey Phipps. 2009. *Fort St. George XII: 2009 excavations at the site of the 1607-1608 Popham colony on the Kennebec River in Maine.* Salem, MA: Peabody Essex Museum.

Brain, Jeffrey Phipps. 2010. *Fort St. George XIII: 2010 excavations at the site of the 1607-1608 Popham colony on the Kennebec River in Maine.* Salem, MA: Peabody Essex Museum.

Brain, Jeffrey Phipps. 2011. *Fort St. George XIV: 2011 excavations at the site of the 1607-1608 Popham colony on the Kennebec River in Maine.* Salem, MA: Peabody Essex Museum.

Brain, Jeffrey Phipps, Ed. 2010. *Popham papers.* Salem, MA: Peabody Essex Museum.

Brewington, M. V. 1962. *Shipcarvers of North America.* Barre: Barre Publishing Company.

Briggs, L. Vernon. 1889. *History of shipbuilding on North River, Plymouth County, Massachusetts, 1640-1872*. Boston: Coburn Brothers.

Briscoe, Mark W. 2005. *Merchant of the Medomak: Stories from Waldoboro Maine's golden years 1860 – 1910*. Waldoboro, ME: Waldoboro Historical Society.

Bunker, Nick. (2010). *Making haste from Babylon: The Mayflower Pilgrims and their world: A new history*. NY: Knopf.

Burrows, Fredrika A. 1974. *Cannonballs & cranberries*. Taunton, MA: William S. Sultwold.

Carter, Matthew. 1997. *The archaeological investigation of a seventeenth-century blacksmith shop at Ferryland, Newfoundland*. M. A. Thesis. St. John's, Newfoundland: Department of Anthropology, Memorial University of Newfoundland.

Carter, Matthew. 1998. A seventeenth-century smithy at Ferryland, Newfoundland. *Avalon Chronicles*. 2: 73-106.

Chaffin, William L. 1886. *History of the town of Easton*. Cambridge, MA: John Wilson and Son.

Chapelle, H. I. 1930. *The Baltimore Clipper, its origin and development*. Salem: Marine Research Society.

Chapelle, Howard I. 1935. *The history of American sailing ships*. New York: Bonanza Books.

Chapelle, Howard I. 1960. *The national watercraft collection*. United States National Museum Bulletin 219. Washington, DC: Smithsonian Institution.

Chard. Jack. 1995. *Making iron and steel: The historic processes, 1700-1900*. Ringwood, NJ: North Jersey Highlands Historical Society.

Chatier, Craig S. 2007. *Archaeology report on South Wareham, Massachusetts*. Unpublished.

Chatterton, E. Keble. 1923. *The mercantile marine*. Boston: Little, Brown & Co.

Cheney, Robert. 1964. *Maritime history of the Merrimac and shipbuilding*. Newburyport, MA: Newburyport Press, Inc.

Crowell, Robert. 1868. *History of the town of Essex: From 1634 to 1868*. Essex, MA: Published by the town.

Currier, John James. 1877. *Shipbuilding on the Merrimac River*. Newburyport, MA: Printed for the author.

Currier, John James. 1909. *History of Newburyport, Mass., 1764-1909*. 3 vols. Newburyport, MA: Printed for the author.

Davis, S. 1815. *Topography and history of Wareham*.

Day, Joan, and R. F. Tylecote, eds. 1991. *The Industrial Revolution in metals*. Brookfield, VT: The Institute of Metals.

Deane, Samuel. 1831. *History of Scituate, Massachusetts, from its first settlement to 1831*. Boston: James Loring.

Demer, John H. 1978. *Jedediah North's tinner's tool business*. South Burlington, VT: The Early American Industries Association.

Deyo, Simeon L. ed. 1890. *History of Barnstable County, Massachusetts*. New York: H. W. Blake & Co.

Diderot, Denis. [1751-75] 1959. *A Diderot pictorial encyclopedia of trades and industry: Manufacturing and the technical arts in plates selected from "L'Encyclopédie, ou Dictionnaire Raisonné des Sciences, des Arts et des Métiers" of Denis Diderot: In two volumes*. Vol. 1 and 2. New York: Dover Publications Inc.

Dieffenbacher, Joseph W. and Dieffenbacher, Jeremy T. (2004). *Midcoast Maine: The Cunningham collection*. Mount Pleasant, SC: Arcadia Publishing.

Doherty, Katherine M. ed. 1976. *History highlights; Bridgewater, Massachusetts a commemorative journal*. Bridgewater, MA: The Bridgewater Bicentennial Commission.

Dow, George Francis. [1927] 1967 facsim. *The arts and crafts in New England: 1704 - 1775*. Da Capo Press Series in Architecture and Decorative Art. vol 1. New York: Da Capo Press.

Dow, George Francis. 1935. *Everyday life in the Massachusetts Bay Colony*. Boston: The Society for the Preservation of New England Antiquities.

Dow, George Francis, and John Henry Edmonds. [1923] 1996. *The pirates of the New England coast 1630 - 1730*. New York: Dover Publications. IS.

Drew, Emily. 1995. *Kingston, the Jones River village as seen by Emily Drew*. Plymouth, MA: Jones River Press.

Dulles, Foster R. 1930. *The old China trade*. Boston: Houghton Mifflin Co.

Dyer, Barbara F. 1984. *Grog ho! The history of wooden vessel building in Camden, Maine*. Camden, ME: B.F. Dyer.

Dyer, Barbara F. 1998. *Vessels of Camden: Images of America*. Mount Pleasant, SC: Arcadia Publishing.

Eaton, Cyrus. 1865. *History of Thomaston, Rockland and South Thomaston, Maine*. vol 1. Hallowell, ME: Masters, Smith & Co.

Eddington, Walter J. 1943. *Glossary of shipbuilding and outfitting terms*. New York: Cornell Maritime Press.

Edwards, William Churchill. 1957. *Historic Quincy, Massachusetts*. Quincy, MA: Published by the city of Quincy.

Elliott, Carrie. 2005. *Life on the river: The flow of Kingston's industries*. Kingston, MA: Town of Kingston and Jones River Village Historical Society.

Ellis, Leonard Bolles. 1892. *History of New Bedford and its vicinity, 1602-1892*. Syracuse, NY: D. Mason & Co.

Elton, G. R. 1967. *England under the Tudors*. London: Methuen & Co. Ltd.

Emery, Samuel Hopkins. 1893. *History of Taunton, Massachusetts: From its settlement to present time*. Syracuse, NY: D. Mason & Co.

Eskew, Garnett Laidlaw. 1958. *Cradle of ships: A history of the Bath Iron Works*. New York: G. P. Putnam's Sons.

Essex Institute. 1934. *The early coastwise and foreign shipping of Salem, 1750-1769*. Salem, MA: The Essex Institute.

Everett, Rev. Noble Warren. 1884. History of Wareham. In *History of Plymouth County, Mass.*, comp by D. Hamilton Hurd. Philadelphia: J. W. Lewis & Co.

Fayle, Charles E. 1933. *A short history of the world's shipping industry*. New York: The Dial Press.

Felt, Joseph B. 1834. *History of Ipswich, Essex, and Hamilton*. Ipswich, MA: Published by the town.

Ferguson, Niall. 2009. *The ascent of money: A financial history of the world*. Penguin Books, NY.

Fitzhugh, William W., and Jacqueline S. Olin, eds. 1993. *Archeology of the Frobisher voyages*. Washington, DC: Smithsonian Institution Press.

French, Gary E. 2010. *Axe making in Ontario in the settlement period*. Elmvale, Ontario, Canada: East Georgian Bay Historical Foundation.

Galer, Gregory J. 2002. Forging ahead: The Ames family of Easton, Massachusetts and two centuries of industrial enterprise, 1635 – 1861. PhD diss., Massachusetts Institute of Technology.

Galer, Gregory, Robert Gordon, and Frances Kemmish. 1999. *Connecticut's Ames Iron Works: Family, community, nature, and innovation in an enterprise of the early American Republic*. New Haven, CT: Connecticut Academy of Arts and Sciences.

Gardner, John. 1975. Mystic Seaport. Personal communications at the Jonesport Wood Co. tool store, West Jonesport, Maine – description of the importance of Josiah Fowler and his Yankee lipped adz for Maine shipwrights.

Garvin, James L. and Garvin, Donna-Belle. (1985). *Instruments of change: New Hampshire hand tools and their makers 1800 - 1900*. Canaan, NH: New Hampshire Historical Society.

Gaynor, Jay. 1993. "Tooles of all sorts to worke": A brief look at common woodworking tools in 17th-century Virginia. In: Reinhart, T. and Pogue, D., eds. *The Archaeology of 17th-century Virginia*. Special Publication no. 30. Courtland, Virginia: Archeological Society of Virginia.

Gaynor, James M., ed. 1997. *Eighteenth-century woodworking tools: Papers presented at a tool symposium: May 19-22, 1994*. Colonial Williamsburg Historic Trades. Volume III. Williamsburg, VA: The Colonial Williamsburg Foundation.

Goldenberg, Joseph A. 1976. *Shipbuilding in colonial America*. Charlottesville, VA: University of Virginia Press.

Goodman, W.L. 1993. *British planemakers from 1700: Third edition*. revised Jane Rees and Mark Rees. Mendham, NJ: Astragal Press.

Gordon, Robert. 1996. *American iron, 1607 - 1900*. Baltimore, MD: Johns Hopkins University Press.

Griffith, Henry S. 1913. *History of the town of Carver, Massachusetts historical review 1637-1910*. New Bedford, MA: E. Anthony & Sons.

Griffith, Richard W. facs. 1867. *The Plymouth County directory and historical register of the Old Colony*. Middleboro, MA: Stillman B. Pratt & Co. http://www.rootsweb.com/~macwareh/Wareham_History_1867.htm

Hain, James. Winter 1994. Woods Hole whaling. *Spritsail: A Journal of the History of Falmouth and Vicinity*. 8: 34-35.

Hall, Henry. 1884. *Report on the ship-building industry of the United States*. Washington: U.S. Bureau of the Census.

Hanley, Michael J. (2011). *Motorcycles, planes, & revolution*. Autumn Woods Studio, Elkhorn, WI.

Hart, William A. 1940. *History of the town of Somerset Massachusetts: Shawomet Purchase 1677, incorporated 1790*. Town of Somerset.

Hartley, Edward Neal. 1957. *Ironworks on the Saugus: The Lynn and Braintree ventures of the company of undertakers of the ironworks in New England*. Norman, OK: University of Oklahoma Press.

Harvey, David. 1985. It's ironmaking time. *Colonial Williamsburg*. 8:9-11.

Harvey, David. 1988. Reconstructing the American bloomery process. In *Historic Trades* vol. 1 *Colonial Williamsburg*. Williamsburg, VA: The Colonial Williamsburg Foundation.

Hayman, Richard. 2005. *Ironmaking: The history and archaeology of the iron industry*. Stroud, UK: Tempus Publishing Ltd.

Heavrin, Charles A. 1998. *The axe and man*. Mendham, NJ: Astragal Press.

Heuvel, Lisa L. 2007. *Early attempts of English mineral exploration in North America: The Jamestown Colony*. Virginia Division of Mineral Resources Publication 176. Charlottesville, VA: Commonwealth of Virginia Department of Mines, Minerals and Energy Division of Mineral Resources.

Hill, Hamilton Andrews. 1892. Thomas Coram in Boston and Taunton. *Proceedings of the American Antiquarian Society* 8: 133-148.

Hindle, Brooke, ed. 1981. *Material culture of the wooden age*. Tarrytown, NY: Sleepy Hollow Press.

Hine, Charles Gilbert. 1908. *The story of Martha's Vineyard, from the lips of its inhabitants, newspaper files and those who have visited its shores, including stray notes on local history and industries*. New York: Hine Brothers.

Honey, Mark E. 2009. *King pine, queen spruce, and jack tar: An intimate history of lumbering on the Union River 1762-1929*. Volume I. Herman, ME: Snowman Printing.

Horne, Charles F. ed. 1894. *Great men and famous women: A series of pen and pencil sketches of the lives of more than 200 of the most prominent personages in history*. New York: Selmar Hess Publisher.

Horsley, John E. 1978. *Tools of the maritime trades*. Camden, ME: International Marine Publishing Company.

Hudson, Alfred Sereno. 1904. *Colonial Concord*. vol. I of *The history of Concord, Massachusetts*. Concord, MA: Erudite Press.

Huffington, Paul, and J. Nelson Clifford. 1939. Evolution of shipbuilding in southeastern Massachusetts. *Economic Geography* 15: 362-378.

Hummel, Charles F. 1968. *With hammer in hand; the Dominy craftsmen of East Hampton, New York*. Published for Henry Francis du Pont Winterthur Museum. Charlottesville, VA: University Press of Virginia.

Huntoon, Daniel T.V. 1893. *History of the town of Canton: Norfolk county Massachusetts*. Cambridge: John Wilson & Son.

Hurd, Duane Hamilton. 1884. *History of Plymouth County, Massachusetts, with biographical sketches of many of its pioneers and prominent men*. Philadelphia: J. W. Lewis & Co.

Hurd, Duane Hamilton. 1888. *History of Essex County, Massachusetts, with biographical sketches of many of its pioneers and prominent men*. Philadelphia: J. W. Lewis & Co.

Hutchins, John G. Brown. 1941. *The American maritime industries and public policy, 1789-1914; an economic history*. Cambridge, MA: Harvard University Press.

Ingraham, Ted. 2010. Plane chatter: Seventeenth-century New England planes. *The Chronicle*. 63: 83-5.

Ingram, David B. 2000. *A brief account of what happened in southeastern Massachusetts after the failure of the old iron works: From early days through the American Revolution*. Ironmasters Conference 2000. Taunton, MA: Old Colony Historical Society.

Ingstad, Anne Stine. 1977. *The discovery of a Norse settlement in America, vol.1, excavations at L'Anse aux Meadows, Newfoundland, 1961-1968*. Oslo, Norway: Universitetsforlaget.

International Correspondence Schools. [1906] 1983. *Blacksmith shop and iron forging*. Bradley, IL: Lindsay Publications Inc.

Jacobs, Donald. 1996. *Bournedale, the forgotten village*. Bourne, MA: The Bourne Historical Commission.

Jewett, Amos Everett, and Emily Mabel Adams Jewett. 1946. *Rowley, Massachusetts: "Mr Ezechi Rogers Plantation" 1639-1850*. Rowley, MA: The Jewett Family of America.

Jones, Henry M. 1926. *Ships of Kingston: "Good-bye, fare ye well"*. Plymouth, MA: The Memorial Press of Plymouth.

Kalman, Bobbie. 2001. *The blacksmith*. NY: Crabtree Publishing Company.

Kauffman, Henry J. 1972. *American Axes: A survey of their development and their makers*. Brattleboro, VT: The Stephen Greene Press.

Keene, Betsey D. 1975. *History of Bourne from 1622 to 1937*. Bourne, MA: Bourne Historical Society.

Kelso, William M. 1996. *Jamestown rediscovery II*. Richmond, VA: The Association for the Preservation of Virginia Antiquities.

Keiler, Hans. 1913. *American shipping, its history and economic conditions*. Jena: G. Fischer.

Kemp, Peter, ed. 1976. *The Oxford companion to ships & the sea*. New York: Oxford University Press.

Klenman, Allen. 1990. *Axe makers of North America*. Victoria, BC: Whistle Punk Books.

Kurlansky, Mark. 2002. *Salt: A world history*. NY: Penguin Books.

Laughton, L. G. Carr. 1925. *Old ship figureheads and sterns*. London: Halton & Truscott Smith, Ltd.

LeBaron, Lemuel, ed. 1907. *Old Rochester and the Old Colony: A collection of data from the New England Magazine*. Dexter, RI.

Lepore, Jill. 1998. *The name of war: King Philip's War and the origins of American identity*. New York: Knopf.

Leonard, Mary Hall. 1907a. Old Rochester and her daughter towns. In *Old Rochester and the Old Colony: A collection of data from the New England Magazine*, ed. Lemuel LeBaron, Dexter, RI.

Leonard, Mary Hall. 1907b. *Mattapoisett and Old Rochester, Massachusetts: Being a history of these towns and also in part of Marion and a portion of Wareham / prepared under the direction of a committee of the town of Mattapoisett*. New York: Grafton Press.

Litchfield, Henry W. 1909. *Ancient landmarks of Pembroke*. Pembroke, MA: G. E. Lewis.

Little, Elizabeth A. 1971. Live oak whaleships. *Nantucket Historic Quarterly*. 19: 24-38.

Loewen, James W. 1995. *Lies my teacher told me: Everything your American history textbook got wrong*. New York: The New Press.

Lovell, R. A. Jr. 1984. *Sandwich: A Cape Cod town*. Sandwich, MA: Sandwich Archives and Historical Center.

Loring, Thelma Rowe. 2001. Historic Bourne: 374 years from the birthplace of commerce in the 1600s to high-tech industries of 2001. *Post Scripts from the Bourne Historical Society*. 15: 3-7.

Lubbock, Basil. 1930. *The Down Easters*. Boston: Charles Lauriat Company.

Lytle, Thomas G. 1984. *Harpoons and other whalecraft*. New Bedford, MA: The Old Dartmouth Historical Society Whaling Museum.

Mack, Edward C. 1949. *Peter Cooper: Citizen of New York*. New York: Duell, Sloan and Pearce.

Madsen, Betsy Ridge, and Maria Burnham. 1981. *Dubbing, hooping, and lofting: Shipbuilding skills*. Manchester, MA: The Cricket Press, Inc.

Marshall, John W. 1888. *History of the town of Rockport: As comprised in the centennial address of Lemuel Gott, M.D., extracts from the memoranda of Ebenezer Pool, Esq., and interesting items from other sources*. Rockport, MA: Rockport Review Office.

Marvin, Winthrop L. 1902. *The American merchant marine*. New York: C. Scribner's Sons.

Mattapoisett Improvement Association. 1932. *Mattapoisett & Old Rochester Massachusetts*. Mattapoisett, MA: The Mattapoisett Improvement Association.

McEntee, Margaret M., Edmund C. Hands, Jeffrey E. Nystrom, Duncan B. Oliver, Hazel L. Varella, and Robert F. Brown, eds. 1974. *History of Easton, Massachusetts: 1886 - 1974*. vol. 2. Easton, MA: Easton Historical Society, Inc.

McKee, Eric. 1972. *Clenched lap or clinker: An appreciation of a boatbuilding technique*. London: National Maritime Museum, Greenwich.

Mendell, Charles S., Jr. 2007. *Mattapoisett sesquicentennial celebration 1857 – 2007 historical summary and souvenir booklet*. Mattapoisett, MA: Mattapoisett Historical Society, Inc.

Merwe, Nikolas J. van der. 1980. The advent of iron in Africa. In *The coming of the age of iron*, Theodore A. Wertime and James D. Muhly, 463 – 506. New Haven: Yale University Press.

Miller, Peggy. 1939. *The New England mind: The seventeenth century*. Cambridge: Harvard University Press.

Mitchell, Nahum. 1970. *History of the early settlement of Bridgewater in Plymouth County, Massachusetts, including an extensive family register*. Baltimore: Gateway Press, Inc.

Morison, Samuel Eliot. 1921. *The maritime history of Massachusetts, 1783-1860, by Samuel Eliot Morison*. New York: Houghton Mifflin Company.

Morison, Samuel Eliot. 1930. *Builders of the bay colony*. Boston: Houghton Mifflin.

Morris, Paul C. 1979. *American sailing coasters of the North Atlantic*. New York: Bonanza Books. IS.

Morrison, Peter H. 2002. *Architecture of the Popham Colony, 1607-1608: An archaeological portrait of English building practice at the moment of settlement*. M.A. Thesis. Orono, ME: University of Maine.

Moxon, Joseph. [1703] 1989. *Mechanick exercises or the doctrine of handiworks*. Morristown, NJ: The Astragal Press.

Muir, Diana. 2000. *Reflections in Bullough's Pond: Economy and ecosystem in New England*. Hanover, NH: University Press of New England.

Mulholland, James A. 1981. *A history of metals in colonial America*. Birmingham, AL: University of Alabama Press.

Murdock, William Bartlett. 1937. *Blast furnaces of Carver: Plymouth County, Massachusetts*. Carver, MA: Carver Historical Commission.

Nason, Elias. 1877. *Gazetteer of Massachusetts*. Boston: B.B. Russell.

Needham, J. 1958. *The development of iron and steel technology in China*. 2 vols. London: Newcomen Society.

Nelson, Robert E., ed. 1999. *Directory of American Toolmakers: A listing of identified makers of tools who worked in Canada and the United States before 1900*. Early American Industries Association.

Nowak, Maryan. Iron making in colonial Taunton. The Taunton River Wild and Scenic Rivers Study. http://www.tauntonriver.org/mills.htm.

Office of the Librarian of Congress. 1882. *The Maine business directory, for 1882.* Boston: Briggs & Co., Publishers.

Old Colony Historical Society. 1901. *A review of the attempt to manufacture iron at Lynn & Braintree in Massachusetts and the successful enterprise at Taunton in the old colony.* Published in the interest of the Memorial to Henry and James Leonard iron masters. Taunton, MA: Executive Committee of the Leonard Family Meeting.

Osgood, Herbert Levi. 1924-25. *The American colonies in the eighteenth century.* 4 vols. New York: Columbia University Press.

Owen, Henry Wilson. 1936. *The Edward Clarence Plummer history of Bath, Maine.* Bath: Times Co.

Pattee, William Samuel. 1878. *A history of old Braintree and Quincy.* Quincy, MA: Green & Prescott.

Packard, Aubigne Lermond. 1950. *A town that went to sea.* Portland, ME: Falmouth Pub. House.

Page, Herb. 2004. *The brothers Coes and their legacy of wrenches.* Davenport, IA: Sunset Mercantile Enterprises.

Pares, Richard. 1956. *Yankees and Creoles: The trade between North America and the West Indies before the American Revolution.* Cambridge: Harvard University Press.

Parks, Edward C. 1857. *The Maine register for 1857 with business directory for the year 1856.* Portland, ME: 82 Exchange Street.

Perley, Sidney. 1924-26. *The history of Salem Massachusetts.* 3 vols. Salem: Sidney Perley.

Perley, Sidney. 2001. *Historic storms of New England.* Beverly, MA: Commonwealth Editions.

Peterson, Harold L. 1956. *Arms and armor of colonial America, 1526-1783.* NY: Bramhall House.

Phillips, James Duncan. 1947. *Salem and the Indies: The story of the great commercial era of the city.* Boston: Houghton Mifflin Company.

Phillips, Stephen Willard. 1934. The ship registers of the District of Newburyport, 1789-1870. *Essex Institute Historical Collections.* LXV.

Pleiner, Radomir. 1962. *Staré evropské kovářství.* Prague: Alteuropäisches Schmiedehandwerk.

Pleiner, Radomir. 1967. *The beginning of the Iron Age in ancient Persia.* Prague: National Technical Museum.

Pleiner, Radomir. 1969. *Iron working in ancient Greece.* Prague: National Technical Museum.

Pleiner, Radomir. 1969. Experimental smelting of steel in early medieval furnaces. *Pamatky archaeologicke.* 458-487.

Pleiner, Radomir. 1979. The technology of three Assyrian artifacts from Khorsabad. *J. Near East. Stud.* 38: 87.

Pleiner, Radomir. 1980. Early iron metallurgy in Europe. In *The Coming of the Age of Iron*, ed. Theodore A.Wertime and James D.Muhly. New Haven: Yale University Press.

Pleiner, R., and Judith K. Bjorkman. 1974. The Assyrian Iron Age: The history of iron in the Assyrian civilization. *Proceedings of the American Philosophical Society.* 118: 283-313.

Pursell, Carroll W. Jr., ed. 1981. *Technology in America: A history of individuals and ideas.* Cambridge, MA: The MIT Press.

Ranbom, Sheppard. 2008. *King Philip's War: A poem.* Arlington, VA: Settlement House.

Raymo, Chet. 2003. *The path: A one-mile walk through the universe.* New York: Walker & Company.

de Réaumur, René Antoine Ferchault. 1722. *L'art de convertir le fer forgé en acier.* Paris, France.

Richardson, M.T., ed. 1978. *Practical blacksmithing: The original classic in one volume.* NY: Weathervane Books.

Ricketson, Daniel. 1858. *The history of New Bedford, Bristol County, Massachusetts: Including a history of the old township of Dartmouth and the present townships of*

Westport, Dartmouth, and Fairhaven, from their settlement to the present time New Bedford Mass. New Bedford: Daniel Ricketson.

Rider, Raymond. 1984. Peek at the past: The story of standard horseshoe. *Wareham Courier*. October 17.

Rider, Raymond. 1989. *Life and times in Wareham over 200 years, 1739-1939.* Wareham, MA: Wareham Historical Society.

Rider, Raymond. 1989. Peek at the past. *Wareham Courier*. June 28. 21.

Rivard, Paul E. 1985. *Made in Maine: An historical overview.* Augusta, ME: Maine State Museum. IS.

Rivard, Paul E. 2007. *Made in Maine: From home and workshop to mill and factory.* Charleston, SC: The History Press.

Roberts, Al, ed. 1970. *Enduring friendships.* Camden, ME: International Marine Publishing Co. IS.

Roberts, Kenneth D. 1976. *Tools for the trades and crafts: An eighteenth century pattern book: R. Timmins & Sons, Birmingham.* Fitzwilliam, NH: Ken Roberts Publishing Co.

Robinson, John, and George Francis Dow. 2007. *Sailing Ships of New England: 1607-1907.* Skyhorse Publishing.

Roubo, André-Jacob. 1761-1777. *L'art due menuisier.* vol 3. Paris: Academie Royale des Sciences.

Rowe, William Hutchinson. 1948. *The maritime history of Maine: Three centuries of shipbuilding & seafaring.* New York: W. W. Norton & Company, Inc.

Ryder, Alice Austin. 1975. *Lands of Sippican on Buzzards Bay.* Marion, MA: Sippican Historical Society.

Sachs, William S., and Ari Hoogenboom. 1965. *The enterprising colonials: Society on the eve of the revolution.* Chicago: Argonaut.

Saltonstall, William Gurdon. 1941. *Ports of Piscataqua; soundings in the maritime history of the Portsmouth, N.H., customs district from the days of Queen Elizabeth and the planting of Strawberry Banke to the times of Abraham Lincoln and the waning of the American clipper*. Cambridge: Harvard University Press.

Sauder, Lee. 1999. The basics of bloomery smelting. *The Anvil's Ring*. 1-6.

Sawyer, Richard P., ed. 1988. *From Pocasset to Cataumet: The origins and growth of a Massachusetts seaside community: Based on the files of Elmer Watson Landers*. Bourne, MA: Bourne Historic Commission.

Sawyer, Richard P. 1998. Beginnings in Bourne: Excerpts from items published in celebration of Bourne's centennial, 1984. *Post Scripts from the Bourne Historical Society*. 12: 3-5.

Schlesinger. 2005. *New York Review of Books*.

Schneider, Paul. 2000. *The enduring shore: A history of Cape Cod, Martha's Vineyard, and Nantucket*. New York: Henry Holt.

Schubert, John Rudolph Theodore. 1957. *The history of the British iron and steel industry from c. 450 B.C. to A.D. 1775*. London: Routledge & Kegan Paul.

Shattuck, Lemuel. 1835. *A history of the town of Concord, Middlesex County, Massachusetts: From its earliest settlement to 1832; and of the adjoining towns, Bedford, Acton, Lincoln, ... and state history not before published*. Boston, Russell, Odiorne and company; Concorde, J. Stacy.

Shaw, John O. 1867. *The Bath, Brunswick and Richmond directory for 1867-8, containing the names of the citizens, and a business directory, with a list of city and town officers, societies, banks, etc.* Boston: Langford & Chase.

Sherby, Oleg D. 1995. Damascus steel and superplasticity – Part I: Background, superplasticity, and genuine Damascus steels. *SAMPE Journal*. 31:10-7.

Sloane, Eric. 1964. *A museum of early American tools*. New York: Henry Holt and Company.

Smith, Cyril S. 1960. *A history of metallography: The development of ideas on the structure of metals before 1890*. Chicago: University of Chicago Press.

Smith, Joseph. [1816] 1975. *Explanation or key, to the various manufactories of Sheffield, with engravings of each article. J. Smith, Sheffield, England.* South Burlington, VT: Early American Industries Association.

Smith, Mary Lou. 1983. *Woods Hole reflections.* Woods Hole, MA: Woods Hole Historical Collection.

Snow, C. P. 1959. *The two cultures and the scientific revolution.* New York: Cambridge University Press.

Spears, John R. 1910. *The story of the American merchant marine.* New York: Macmillan.

Spring, Laverne W. 1992. *Non-technical chats on iron and steel and their application to modern industry.* Bradley, IL: Lindsay Publications Inc.

St. George, Robert Blair. 1979. *The wrought covenant: Source material for the study of craftsmen and community in southeastern New England: 1620-1700.* Brockton, MA: Brockton Art Center and Fuller Memorial.

Stahl, Jasper Jacob. 1956. *History of Old Broad Bay and Waldoboro: Volume one: The colonial and federal periods.* Portland, ME: The Bond Wheelwright Company.

Starbuck, Alexander. [1878] 1964. *History of the American whale fishery from its earliest inception to the year 1876.* 2 vols. New York: Argosy-Antiquarian Ltd.

Stewart, John, and Henry Unglik. 1999. *Metallurgical investigation of archaeological material of Norse origin from L'anse aux Meadows, Newfoundland.* Ottawa: Parks Canada.

Story, Dana. 1995. *The shipbuilders of Essex: A chronicle of Yankee endeavor.* Gloucester, MA: Ten Pound Island Book Company.

Survey of Federal Archives and The National Archives. 1940. *Ship registers of New Bedford, Massachusetts: Volume I: 1796 - 1850.* Boston: The National Archives Project.

Taussig, F.W. 1931. *Tariff history of the United States.* New York: G.P. Putnam's Sons.

Thayer, Henry O., ed. [1892] 1970. *The Sagadahoc Colony comprising the relations of a voyage into New England.* NY: Research Reprints.

Thayer, Philip S. 1984. The 1669 will and inventory of Michael Willis of Boston, cutler. *The Chronicle*. 37:61-63.

Thompson, Edward V. 2010. *Printed maps of the district and state of Maine 1739-1860: An illustrated and comparative study.* Bangor, ME: Nimue Books & Prints.

Thompson, Elroy S. 1928. *History of Plymouth, Norfolk, and Barnstable counties Massachusetts.* 3 vols. New York.

Tremblay, Robert and David-Thiery Ruddel. 2010. *"By hammer and hand all arts do stand: Blacksmithing in Canada before 1950.* Ottawa, Canada: Canadian Science and Technology Museums Corporation.

Tweedale, Geoffrey. 1983. *Sheffield steel and America: Aspects of the Atlantic migration of special steelmaking technology, 1850-1930.* London: F. Cass & Co.

Tylecote, Ronald F. (1962). *Metallurgy in archaeology; a prehistory of metallurgy in the British Isles.* London: Edward Arnold.

Tylecote, Ronald F. 1987. *The early history of metallurgy in Europe.* London: Longmans Green. IS.

Walcott, Charles. 1884. *Concord in the colonial period: Being a history of the town of Concord, Massachusetts, from the earliest settlement to the overthrow of the Andros government, 1635-1689.* Boston: Estes and Lauriat.

Waters, Henry. 1982. *Kings' Mills: Whitefield, Maine 1772-1982.*

Wayman, Michael L. 2000. *The ferrous metallurgy of early clocks and watches: Studies in post medieval steel.* Occasional Paper Number 136. London: British Museum.

Wertime, Theodore A. 1962. *The coming of the age of steel.* Chicago: University of Chicago Press. IS.

Wertime, Theodore A., and James D. Muhly, ed. 1980. *The coming of the age of iron.* New Haven: Yale University Press. IS.

Weston, Thomas. 1906. *History of the town of Middleboro Massachusetts.* New York: Houghton, Mifflin and Company.

Wilson, Daniel Munro. 1926. *Three hundred years of Quincy, 1625-1925, historical retrospect of Mount Wollaston, Braintree, and Quincy, by Daniel Munro Wilson*. Quincy, MA: Published by authority of the city government of Quincy.

Wilson, Margaret Jerram. 2007. *Norumbega navigators: Early English voyages to New England and the story of the Popham Colony*. Bath, England: Wilson Publications. IS.

Wing, Anne, and Donald Wing. 2005. *Early planemakers of London: Recent discoveries in the Tallow Chandlers and the Joiners Companies*. Marion, MA: The Mechanik's Workbench.

Wing, Donald, and Anne Wing. 1990. The Pool family of Easton, Massachusetts. *Rittenhouse: Journal of the American Scientific Instrument Enterprise*. 4: 118-126.

Winthrop, John Sr. *Winthrop papers*. Boston: Massachusetts Historical Society Collections 5th series, VIII.

Wood, Edward F. R. Jr., and Judith Navas Lund. 2004. *The ports of old Rochester: Shipbuilding at Mattapoisett and Marion*. Mattapoisett, MA: New Bedford Whaling Museum and Quadequina Publishers.

Made in the USA
Middletown, DE
11 July 2021